CAMBRIDGE LIBRARY COLLECTION

Books of enduring scholarly value

Physical Sciences

From ancient times, humans have tried to understand the workings of
the world around them. The roots of modern physical science go back to
the very earliest mechanical devices such as levers and rollers, the mixing
of paints and dyes, and the importance of the heavenly bodies in early
religious observance and navigation. The physical sciences as we know them
today began to emerge as independent academic subjects during the early
modern period, in the work of Newton and other 'natural philosophers',
and numerous sub-disciplines developed during the centuries that followed.
This part of the Cambridge Library Collection is devoted to landmark
publications in this area which will be of interest to historians of science
concerned with individual scientists, particular discoveries, and advances in
scientific method, or with the establishment and development of scientific
institutions around the world.

A New System of Chemical Philosophy

The renowned English chemist and meteorologist John Dalton (1766–1844)
published *A New System of Chemical Philosophy* in two volumes, between
1808 and 1827. Dalton's discovery of the importance of the relative weight
and structure of particles of a compound for explaining chemical reactions
transformed atomic theory and laid the basis for much of what is modern
chemistry. Volume 2 was published in 1827. It contains sections examining
the weights and structures of two-element compounds in five different
groups: metallic oxides; earthly, alkaline and metallic sulphurets; earthly,
alkaline and metallic phosphurets; carburet; and metallic alloys. An appendix
contains a selection of brief notes and tables, including a new table of the
relative weights of atoms. A planned second part was never published.
Dalton's work is a monument of nineteenth-century chemistry. It will
continue to be read and enjoyed by anybody interested in the history and
development of science.

Cambridge University Press has long been a pioneer in the reissuing of out-of-print titles from its own backlist, producing digital reprints of books that are still sought after by scholars and students but could not be reprinted economically using traditional technology. The Cambridge Library Collection extends this activity to a wider range of books which are still of importance to researchers and professionals, either for the source material they contain, or as landmarks in the history of their academic discipline.

Drawing from the world-renowned collections in the Cambridge University Library, and guided by the advice of experts in each subject area, Cambridge University Press is using state-of-the-art scanning machines in its own Printing House to capture the content of each book selected for inclusion. The files are processed to give a consistently clear, crisp image, and the books finished to the high quality standard for which the Press is recognised around the world. The latest print-on-demand technology ensures that the books will remain available indefinitely, and that orders for single or multiple copies can quickly be supplied.

The Cambridge Library Collection will bring back to life books of enduring scholarly value (including out-of-copyright works originally issued by other publishers) across a wide range of disciplines in the humanities and social sciences and in science and technology.

A New System of
Chemical Philosophy

VOLUME 2

JOHN DALTON

CAMBRIDGE UNIVERSITY PRESS

Cambridge, New York, Melbourne, Madrid, Cape Town, Singapore,
São Paolo, Delhi, Dubai, Tokyo

Published in the United States of America by Cambridge University Press, New York

www.cambridge.org
Information on this title: www.cambridge.org/9781108019682

© in this compilation Cambridge University Press 2010

This edition first published 1827
This digitally printed version 2010

ISBN 978-1-108-01968-2 Paperback

A

NEW SYSTEM

OF

CHEMICAL PHILOSOPHY.

PART FIRST

OF

VOL. II.

BY

JOHN DALTON, F.R.S.

President of the Literary and Philosophical Society, Manchester;
Corresponding Member of the Royal Academy of Sciences, Paris;
Member of the Royal Academy, Munich, and of the Cæsarean
Natural History Society, Moscow;
Honorary Member of the Royal Medical Society, Edinburgh,
and of the Philosophical Societies of Bristol, Cambridge,
Leeds, Sheffield and Yorkshire.

Manchester:
Printed by the Executors of S. Russell,
FOR
GEORGE WILSON, ESSEX STREET, STRAND,
LONDON.

1827.

PREFACE.

THE work now submitted to the public was begun to be printed in 1817; and the 13th and 14th sections, containing the oxides and sulphurets, were printed off before the end of October of the same year. The printing of the rest of the work to the appendix was finished in September, 1821. One sheet of the appendix was printed at the end of 1823; but no addition was afterwards made till May, 1826; when the printing was resumed, and has been continued to the present time.

It may be asked, what were the motives for such a plan of procedure. To this it may be replied, that soon after the publication of the first volume (in 1810), I began to prepare materials, and to institute experiments, relating to the oxides, &c., with occasional diversions into other departments of chemistry, as circumstances arose. As a great portion of my time was always necessarily engaged in professional duties, and as that part of the work I was about to commence was one running into detail, I thought it would be best to print it as I proceeded, whilst the train of thought and of experiments was fresh in view. The advantage in this case was expected to be partly at least counterbalanced by the loss of discoveries and improvements likely to he made in the interval between the printing and publishing of the several articles. This I was aware of; but as a principal object I had in view was to

viii.   PREFACE.

give the results of my own experience, in the various departments of chemical science, rather than to form the best compilation of Chemistry at the period, this object was most likely to be obtained by the proposed plan. It is true the time the work has been in the press has far exceeded my expectation; notwithstanding this I am not conscious of any very material alterations or additions, which I should wish to make at the present moment.

It affords me great pleasure to acknowledge the assistance I have had during the progress of this volume, from a valuable selection of chemical apparatus, for which I am indebted to the generosity of Mr. Sharpe; also the continued and friendly intercourse with Dr. Henry, whose discussions on scientific subjects are always instructive, and whose stores are always open when the promotion of science is the object.

My present design is to add a second part to this volume, and with that to finish the work. It will consist of the more complex compounds. Acids, and other products of the vegetable kingdom, Salts, &c., will form principal parts. Already I have a stock of experiments on these subjects; but I am not satisfied without exploring this region afresh.

August, 1827.

CONTENTS OF VOL. II.

Part First.

Page.

CHAP. V.—COMPOUNDS OF TWO ELEMENTS.

SECTION 13. *Metallic Oxides*................................ 1
Oxide of Gold........................... 5
———— *Platina* 11
———— *Silver* 17
Oxides of Mercury 19
Oxide of Palladium 24
Oxides of Rhodium, Iridium, and
Osmium. 26
———— *Copper*...................... 26
———— *Iron* 28
———— *Nickel* 34
———— *Tin* 36
———— *Lead*............................ 39
Oxide of Zinc 51
Oxides of Potassium.................... 53
———— *Sodium* 56
Oxide of Bismuth 57
Oxides of Antimony..................... 58
Oxide of Tellurium 62
Oxides of Arsenic....................... 63
———— *Cobalt* 68
———— *Manganese* 71
———— *Chromium* 80
———— *Uranium* 86
———— *Molybdenum*.................. 87
———— *Tungsten* 90
———— *Titanium* 91
———— *Columbium* 92
———— *Cerium*....................... 94

Page

SECTION 14 *Earthy, Alkaline, and Metallic Sul-*
                         phurets 96

        Sulphurets of Lime 99

        Sulphuret of Magnesia 111

        Sulphurets of Barytes 112

        ———— *Strontites* 114

        ———— *Alumine, Silex, Yttria,*
                    Glucine and Zircone ... 114

        ———— *Potash* 116

        ———— *Soda* 119

        Sulphuret of Ammonia 120

        Sulphurets of Gold 121

        Sulphuret of Platina 123

        Sulphurets of Silver 126

        ———— *Mercury* 127

        Sulphuret of Palladium 131

        ———— *Rhodium* 132

        ———— *Iridium* 132

        ———— *Osmium* 132

        Sulphurets of Copper 133

        ———— *Iron* 134

        ———— *Nickel* 138

        ———— *Tin* 139

        ———— *Lead* 144

        ———— *Zinc* 146

        ———— *Potassium and Sodium* ... 148

        ———— *Bismuth* 149

        ———— *Antimony* 151

        Sulphuret of Tellurium 153

        Sulphurets of Arsenic 153

        Sulphuret of Cobalt 160

        Sulphurets of Manganese 162

        Sulphuret of Chromium 163

        ———— *Uranium* 164

        ———— *Molybdenum* 164

        Sulphuret of Tungsten 164

Page.

Sulphurets of Titanium, Columbium, and Cerium 165

SECTION 15. *Earthy, Alkaline, and Metallic Phos-*
      phurets...................... 166

  Phosphuret of Hydrogen 169

  Phosphurets of Carbon and Sulphur 184

  Phosphuret of Lime...... 184

  ————— *Barytes* 188

  ————— *Strontites*.................... 190

  ————— *Gold* 191

  ————— *Platina* 194

  ————— *Silver* 195

  ————— *Mercury* 197

  ————— *Palladium* 198

  ————— *Copper*...................... 199

  ————— *Iron*...................... .. 201

  ————— *Nickel* 201

  ————— *Tin* 202

  ————— *Lead* 203

  Phosphurets of Zinc and Potassium 204

  ————— *Sodium and Bismuth* 207

  ————— *Antimony and Arsenic* ... 208

  Phosphuret of Cobalt 209

  ————— *Manganese* 210

SECTION 16. *Carburets*........... 211

  ——— *of Iron...steel* ....212—214

SECTION 17. *Metallic Alloys*........................... 218

  Alloys of Gold, with other metals.... 222

  ————— *Platina, with other metals* 226

  ————— *Silver, with other metals*... 228

  ————— *Mercury, and other metals .*
      amalgams.................. 230

  Triple, Quadruple, &c. amalgams ... 236

  Alloys of Copper, with other metals 238

  ————— *Iron, with other metals* ... 253

  Alloys of Nickel and Tin, with do.... 254

  ————— *Lead with do.* 258

                                                     Page.
SECTION 17. *Triple Alloys, Solders ; fusible metal, &c.* 263

                    APPENDIX.

Abstract of De la Roche and Berard's essay on the
          specific heat of gases 268
———— *Dulong and Petit's essays,*
      On the expansion of air, mercury, glass, iron,
          copper, and platina, by heat 272
      On the capacities of certain bodies, for heat 274
      On the laws of refrigeration 277
      On the specific heats of certain bodies 280
    Remarks on the above essays 282
New Table of the forces of vapours 298
Table of the expansion of air, and the force of aqueous
          and ætherial vapour, adapted to atmospheric
          temperatures............................ 299
Applications of the above table 300
Formulæ for determining the proportions of combustible
          gases, in mixtures 305
Heat produced by the combustion of gases............... 309
Absorption of gases by water 309
Fluoric acid—deutoxide of hydrogen 311
Muriatic acid—oxymuriatic acid 313
Nitric-acid—compounds of azote and oxygen..... 315
On ammonia..... 328
Decomposition of ammonia by nitrous oxide 330
——————————————— *by nitrous gas and oxygen* 332
Volume of gases-from the decomposition of ammonia... 335
Decomposition of ammonia by a red heat 335
Decomposition of ammonia by oxymuriatic acid......... 335
Sulphuret of Carbon 338
Potassium, Sodium, &c 340
Alum 341
New table of the relative weights of atoms. 352
ADDENDA. *Steel ; mixed gases ; expansion of liquids*
          by heat 354

NEW SYSTEM

OF

CHEMICAL PHILOSOPHY.

CHAP. V.

SECTION 13.

METALLIC OXIDES.

ALL the metals are disposed to combine with oxygen, but the combination is effected more easily with some than with others; the compound is usually called an *oxide*, but in some instances it is also called an *acid*. The same metal combines with one, two, or perhaps more atoms of oxygen, forming compounds which may be distinguished according to Dr. Thomson, by the terms *protoxide, deutoxide, tritoxide,* &c.

Such however is the repulsion of oxygen to oxygen that we rarely find three atoms of it retained by a single atom of any kind; and there are not many instances of metals capable of holding two atoms of oxygen. Various modifications of the proportions of metals and oxygen arise from the combinations of the oxides themselves one with another and with oxygen, so as to lead some to imagine that an atom of metal in some instances combines with 3, 4, or more of oxygen. This is altogether improbable: It is much more simple to suppose that one atom of oxygen connects two or more atoms of protoxide, 1 of protoxide unites to 1 or more of deutoxide, &c. These intermediate oxides are in few if any instances found to combine with acids like the other two oxides.

There is no reason that I am acquainted with for disbelieving that oxygen combined with a metal is still repulsive of oxygen, and that by the same law as particles of an elastic fluid; that is, the repulsion is inversely as the distance of the centres of the atoms. Hence it may be demonstrated that it requires twice the strength of affinity to form a deutoxide as a protoxide, three times the strength to form a tritoxide as a protoxide, &c. On this account it is, in all probability, that deutoxides are not numerous, and tritoxides are rarely if ever found.

The quantity of oxygen that combines

with any metal to form an oxide may be investigated by several methods.

1st. By combustion; a given weight of the metal may be burned and the oxide produced may be collected and weighed; when the increase by combustion will appear.

2. By solution in an acid and precipitation by an earth or alkali; in this case a given weight of the metal is dissolved and precipitated; the precipitate collected and sufficiently dried shews the increase by oxygen.

3. By transferring the oxygen from an oxide to another metal; in this case the metal in question is usually immersed in a saline solution of the other metal; this latter metal gives up its oxygen to the former and is itself reformed or *revived* as it is termed.

4. By determining the proportion of hydrogen gas evolved during the solution of a given weight of metal; then allowing half of that volume for its equivalent of oxygenous gas, the weight of it shews the oxygen united to the metal; it being now well understood that water furnishes the two elements of hydrogen and oxygen in such case.

5. The higher oxides are conveniently determined by the application of the solution of oxymuriate of lime to the lower oxides in solution.

6. The quantity of oxygen in several oxides may be found from the quantity of nitrous gas evolved during the solution of a given weight of metal in nitric acid.

The first four methods have been used by chemists for several years past; the two last I have added from my own experience, having found them very useful assistants in various instances. The last method by nitrous gas, has indeed been proposed before, and labour bestowed on it both by others and myself, but without reducing the results to any certainty, till lately; the principal cause of this want of success has arisen from misunderstanding the nature and constitution of nitric acid. Most chemists seem with me to have mistaken *nitrous* acid for *nitric;* the former is composed of 1 atom of azote and 2 of oxygen; or perhaps of 2 azote and 4 oxygen; the latter of 2 azote and 5 oxygen, or 2 nitrous gas and 3 oxygen; the weight of the former is 19, or its double 38, on my scale, and that of the latter 45. [My reasons for adopting the above conclusion respecting nitrous acid, which is at variance with that in VOL. 1, p. 331, will be given hereafter.] When therefore a metal is oxidized by nitric acid, 3 atoms of oxygen $(= 21)$ go to the metal, and 2 atoms of nitrous gas $(= 24)$ are disengaged.

Hence $\frac{7}{8}$ of the weight of nitrous gas evolved is the weight of oxygen combined. It sometimes happens however that the nitrous gas is partly or wholly retained by the residue of nitric acid; but in this case the oxymuriate of lime can be applied to convert the nitrous gas into nitric acid, and from the oxygen imbibed the quantity of nitrous gas may be inferred.

1. *Oxide of Gold.*

Some difficulties have been found in ascertaining both the number and proportions of the oxides of gold; hence the differences in the results of authors.

Gold does not burn by exposure to heat, but gold leaf and gold wire may be deflagrated by electricity and galvanism; a purple powder is the product, which is considered by some as the protoxide of gold; but others, after Macquer and Proust, conceive with greater probability that this powder is nothing but gold reduced to its ultimate division. Solutions of gold which are of a fine yellow, give a purple stain; and gold deoxidized by green sulphate of iron is precipitated blue, which precipitate gradually assumes a yellow colour as the particles become

united. The very weak affinity of gold for
oxygen is shewn by the difficulty with which
it is oxidized and the ease with which the
oxygen is expelled again by heat; these
facts seem to preclude the idea of gold combining with oxygen in high temperatures.

Protoxide. Gold is scarcely affected by
pure sulphuric, nitric or muriatic acid; but
it is easily oxidized and dissolved by nitro-
muriatic acid (that is, a mixture of nitric
and muriatic acids) when assisted by a temperature of 150 or 200°. Caustic potash being put into the solution and heated, a brownish black precipitate is obtained; but a part
of the oxide remains in solution combined
with the muriate of potash, according to
Vauquelin; and Proust has observed that the
oxide cannot be washed and dried in a moderate heat without a portion of the gold being
revived; hence the difficulty of ascertaining in this way the weight of oxygen combining with gold.

I have succeeded, as I apprehend, in determining the relative weights of gold and
oxygen, by two methods, which mutually
corroborate each other. The first is by
means of the nitrous gas generated by the
solution of gold; and the second is, by finding what quantity of green oxide of iron is

converted into red by precipitating a given weight of gold in solution.

Ten grains of guinea gold of the sp. gr. 17.3, were repeatedly dissolved in a small excess of nitro-muriatic acid; the quantity and purity of the nitrous gas generated were duly observed and allowance made for the loss occasioned by a small portion of common air originally in the gas bottle. The volume of nitrous gas corrected as above was always found between 1100 and 1200 grain measures, the weight of which may be estimated at 1.6 grains, corresponding to 1.4 grains of oxygen. The small portion of alloy $(\frac{1}{12})$ known to be in standard gold is chiefly copper with a small part silver; now it will be seen in the sequel that copper takes $\frac{1}{4}$ of its weight of oxygen ; hence if we allow .8 of a grain for copper and .2 for the oxygen combining with it, we shall have 9.2 gold united to 1.2 oxygen, or 100 gold with 13 oxygen, which is nearly the same as Berzelius has determined by precipitating the gold by mercury.—Again, 10 grains of gold were dissolved as above (= 9.2 pure) and precipitated by a solution of pure green sulphate of iron of the sp. gr. 1.181 and which I had previously proved to contain 9 grains of green oxide in 100 measures. They converted 120 measures of this green

sulphate into yellow, which was carefully
precipitated afterwards by lime water, dried
and weighed. The gold precipitated was
found very nearly 9 grains; and the yellow
oxide of iron mixed with oxide of copper
was nearly 13 grains. Now 120 measures
sulphate iron contain 10.8 grains green oxide,
and these require $\frac{1}{9}$ of their weight of oxygen (see the oxides of iron) to be changed
into yellow oxide, or 1.2 oxygen. Hence it
appears that the oxygen combined with the
gold was transferred to the iron unchanged in
quantity. It is to be observed however that
green oxide of iron not only deoxidates the
gold but it semideoxidates the copper also;
so that .1 of the transferred oxygen above
might be said to be derived from the copper,
and the rest, or 1.1 from the 9 grains of gold;
this would give 100 gold to 12.2 oxygen,
which is still nearer to the determination of
Berzelius. Upon the whole I am inclined to
adopt the proportion of 8 to 1 or 100 to 12.5
as that which appears the most correct approximation and at the same time a ratio
easily remembered and adapted to facilitate
calculations.

We are now to consider whether the above
is the protoxide. As no other oxide has been
clearly shewn to exist, and as this combines

with muriatic acid, with ammonia, with
the oxide of tin, &c. and is wholly deoxi-
dated by green sulphate of iron and by a moderate heat, there seems every reason to conclude it is a combination of the most simple
kind, or 1 atom of metal to 1 of oxygen.
Hence the atom of oxygen being 7, that of
gold must be 56, and not 140 or 200, as stated VOL. 1, p. 250.

Berzelius seems to consider the above as the
tritoxide, or three atoms of oxygen to one of
gold; but it is extremely improbable that
gold, which is allowed to have a weak affinity
for oxygen, should be able to restrain the
violent repulsion of three atoms of oxygen,
and should on every occasion lose them all at
once, and not by degrees, as is usual with
other high oxides.

Subjoined are the results of various authors in regard to the oxide of gold, but generally given with diffidence as to their accuracy.

	gold	oxygen
Bergman	100	+ 10
Proust	——	+ 8.57 to 31.
Oberkampf	——	+ 10
Berzelius	——	+ 12 (4, suboxide)
My results	——	+ 12.5

Since writing the above I have had an opportunity of repeating the experiments on the oxide of gold by an improved nitrous gas apparatus, calculated almost entirely to exclude atmospheric air; I find less nitrous gas produced during the solution than stated above, sometimes by $\frac{1}{7}$, and that it is variable according to the excess of nitric acid; also that the solution requires a portion of oxymuriatic acid as an equivalent for the nitrous gas retained. I prefer, however, the method of oxidizing the green sulphate of iron; by putting a small excess of the green sulphate and precipitating, first the red oxide and then the green, I obtained very distinct results. On the whole I am inclined to think my results preceding these have rather overrated the oxygen, and that it would as nearly be stated at 11 on the hundred. This would be nearly a mean of those in the above table, and would require the atom of gold to be 63, and that of the oxide 70. Between the two extremes of 56 and 63 it is most probable the true weight of the atom of gold will be found.

It may be proper to add that I have found 100 grain measures of muriatic acid (1.16), and 25 of nitric (1.35), are sufficient to dissolve 40 grains of standard gold; and I have reason to think the acids are in due proportion nearly, though different from what is usually recommended and employed.

2. Oxide of Platina.

Platina exhibits greater difficulties than gold in the investigation of its compounds with oxygen. It is not oxidized by heat; but by the explosion of an electric battery it is converted into a black powder, which is most probably the metal in extreme division, though it has been considered by some as the protoxide. Platina is capable of being oxidized and dissolved by nitro-muriatic acid, but less easily than gold; it requires more acid, as high or higher temperature and long continued digestion; nitrous gas is given out, during the solution, but sparingly. When lime or an alkali is added to the solution with a view to precipitate the oxide, a triple compound is usually formed of the acid, the oxide and the alkali, which is in most instances precipitated. This weighty compound renders the valuation of the oxygen in it very uncertain.

Chenevix has made some observations on the oxides of platina, (see Nichols. Journ. 7. p. 178.) He finds two oxides: the one consists of 93 platina and 7 oxygen; the other of 87 platina and 13 oxygen; but the experiments on which these results rest are not quite satisfactory.

Mr. E. Davy in the 40th vol. of the Philos. Magazine, states his having reduced the oxide of platina in solution by means of hydrogen; and that he finds the oxide to consist of 84 platina and 16 oxygen nearly. I could not succeed at all in effecting the reduction of the metal in this way.

Berzelius has lately given us the results of his investigation on this subject. (An. de Chimie 87—126.) According to this distinguished chemist there are two oxides of platina; the first consists of 100 metal and $8\frac{1}{4}$ oxygen, and the second of 100 metal and $16\frac{1}{2}$ oxygen, nearly. In order to understand his process it may be proper to premise that when nitro-muriatic acid has dissolved as much platina as it can, there is still a great excess of one or both of the acids, which is unnecessary for the existence of the solution, and which may, and in general ought to be expelled by evaporation; by exposing the solution to a heat of 100 or 150° the excess of both acids is in great part driven off and a dry red mass is obtained, without smell, but very deliquescent. It is equal to or rather more than twice

the weight of the platina. It consists of water, muriatic and nitric acids, oxygen and platina; it is still an acid salt. By exposing the
dry mass again to a heat of 400 or 500°, it
liquifies, exhales acid fumes having the odour
of oxymuriatic acid, and becomes again a
dry mass of an olive colour, exhaling fumes as
the heat increases, and loses about ¼ of its
weight. It is still soluble in water, except a
few atoms of black powder, continues acid
to the tests, and may be considered as a supermuriate of platina. If this olive powder
be again heated almost to red, it exhales a visible smoke in the open air, which has the
smell of oxymuriatic acid, and becomes a
light brown powder, having lost a little
weight. It is then neither deliquescent nor
soluble in water except in a small degree so
as to give the yellow colour. In this state it
has been considered as a neutral muriate.
By a moderately bright red heat this powder
is decomposed and leaves a black spongy
mass which is found to be pure platina.

The insoluble muriate of platina according
to Mr. E. Davy, contains 72.5 per cent. of
platina, and Berzelius finds 73.3; the loss is
considered as oxymuriatic acid; hence from
the known proportions of this acid Berzelius
infers the constituents of 100 muriate=73.3

platina, 6.075 oxygen and 20.625 muriatic acid; or 100 platina take 8.3 oxygen. The near agreement of the above authors is favourable to the accuracy of their results; but I have found some difficulty in obtaining the insoluble muriate free from the soluble one, and at the same time from reduced platina because the precise degree of heat requisite to produce it is neither well known nor easily attained; and it is desirable that a certain weight of platina should be dissolved and the same weight reproduced as a confirmation of accuracy. From a train of experiments on the soluble and insoluble muriates of platina, the salts being obtained from the purified laminated metal, I am disposed to consider the above results as good approximations to the truth.

In order to obtain the other oxide, Berzelius digests mercury in a solution of the supermuriate of platina; a black powder is thrown down, which is found to be platina, and mercury is taken up, being oxidized at the expence of the platina. It was found that 16.7 grains of mercury had precipitated 8.5 of platina; and the mercury being calculated as in the state of deutoxide, contained, from the known proportions of this metal, 1.4 oxygen; hence 8.5 platina must have yielded 1.4 oxygen;

and if 8.5:1.4:: 100:16.4; so that 100 platina must have had 16.4 oxygen in the supermuriate, or twice the quantity it had in the insoluble muriate.

This conclusion appears to me premature; the mercurial oxide should at least have been precipitated and a corresponding quantity have been found and proved to be the red oxide. Even had this been the case, it is not easy to determine what quantity of it might be due to the residue of nitro-muriatic acid. But I have not found that the common yellow or red oxide of mercury is precipitated by lime water in such case; the precipitate is brown, and evidently contains both mercury and platina. Proust had found in his excellent essay on platina (Journ. de Physique 52—437, 1801) that mercury decomposes muriate of platina, that an amalgam of platina with a little calomel, and much mercury in powder, were precipitated; exposed to heat, a fine black powder was left which had the characters of platina. Into a solution of pure platina that had been evaporated to dryness in 150° and redissolved, I put 9¼ grs. of mercury, and boiled it for 10 minutes in a glass capsule, till there was apparently no further change; the liquor filtered was as yellow as at first; the mixture of black powder and running mercury remaining on the filter, when dried, weighed 6½ grains; this heated to a low red in an iron spoon, left 1 grain of fine black powder; the liquid saturated with lime water, yielded 2½ grains dry black powder insoluble in cold nitric acid; after this, protomuriate of tin threw down 5¾ grains of the compound oxides of platina and tin. The solution at first contained 3.3 grains of platina.

In another experiment 2 parts of calomel were put to 1 of platina in solution; when heated to boiling, the calomel was dissolved and a little black powder was precipitated, which did not amount to half the weight of the platina. Lime water threw down from the liquid, a yellowish olive or brown precipitate, partially soluble in cold nitro-muriatic acid; and after this, muriate of tin yielded a brown precipitate. These experiments shew that the action between muriate of platina and mercury or the mercurial salts, is of a complicated nature, and is not limited to the decomposition of the oxide of platina and the substitution of the deutoxide of mercury in its place.

The difficulties abovementioned have led me to investigate the oxygen combining with platina by means of the nitrous gas yielded upon its solution in nitro-muriatic acid. By three distinct experiments I found that 10 grains of pure platina by solution yielded nearly 750 grain measures of nitrous gas, which may be considered as 1 grain in weight ; $\frac{7}{8}$ of which $= .875$ for oxygen; this would give 8.75 oxygen per cent. But from a subsequent experiment made under circumstances calculated to preclude as much as possible every source of fallacy, I obtained 790 measures of nitrous gas from 10 grains of platina; and the solution afterwards took 60 measures of oxymuriatic acid gas before a permanent smell of the gas was produced. These 790 measures $= 1.05$ grain, $\frac{7}{8}$ of which $= .92$, to which add .04 for the oxygen acquired from the oxymuriatic acid, and we have ,96 oxygen for 10 platina ; or 100 platina take 9.6 oxygen. But if $9.6 : 100 :: 7 : 73$, for the weight of an atom of platina, and 80 for that of the protoxide, as I apprehend it to be, and the only oxide of platina we can at present form. (The atom of platina in Vol. 1, page 248, was estimated at 100.)

3. *Oxide of Silver.*

When silver wire is exploded by electricity in oxygen gas, a black powder is produced, which is the oxide of silver. If silver be dis-

solved in nitric acid and precipitated by lime
water, an olive brown powder falls which becomes black when exposed to the light. This
black powder is the only oxide of silver with
which we are acquainted. The proportion of
silver and oxygen has been investigated by
various chemists; the results are as under.

	silver		oxygen
Wenzel..................	100	+	8.5
Proust.............. ...		+	9.5
Bucholz and Rose ...		+	9.5 *
Gay Lussac............		+	7.6 †
Berzelius..............		+	7.925

From the solution of 170 grains of standard
silver I obtained nearly 30 oz. measures of nitrous gas $= 18\frac{1}{2}$ grains, corresponding to 16
oxygen. This would give 9.4 oxygen upon
100 silver. But as $\frac{1}{10}$ of the metal or 17
grains was copper, and this takes $\frac{1}{4}$ of its
weight of oxygen, we shall have 159 silver
and $11\frac{3}{4}$ oxygen, or 100 silver and 7.7 oxygen
nearly.

If we adopt 7.8 as the proper quantity of
oxygen on 100 silver, we shall have 7.8 :
100 :: 7 : 90 nearly, which represents the
weight of an atom of silver, and 97 that of
the oxide.

* 7.9 when duly corrected. Annal. de Chimie, 78—114.
† Memoirs d'Arcueil 2—168.

4. *Oxides of Mercury.*

Two oxides of mercury have been long known and are well distinguished from each other. They may be obtained by exposing mercury to a heat not exceeding 600°, in contact with oxygen gas or atmospheric air, and due agitation; but this method is rarely adopted in practice. A high degree of heat decomposes the oxides again.

Protoxide. To obtain the protoxide, mercury must be slowly dissolved in dilute nitric acid without heat, and an excess of mercury must be used. If to 1000 grain measures of nitric acid, 1.2 sp.gr. be put 500 grains of mercury, by occasional agitation the requisite solution will be obtained in 24 hours. A portion of this solution must be treated with a small excess of lime water or caustic alkali, when a black powder will be thrown down, which is the oxide containing a minimum of oxygen, and hence may be considered the protoxide.

Deutoxide. If to 1000 measures of nitric acid, 1.2 sp. gr. be put 350 grains of mercury, and the mixture be boiled till the mercury disappear, a solution will be obtained containing the deutoxide. A portion of this being treated as beforementioned with lime water, a yellowish red powder is precipitated, which is the oxide of mercury containing a maximum of oxygen; all the later authors agree that it contains just double the quantity of oxygen to a given portion of mercury that the former does, and may therefore be called the deutoxide.

These two oxides combine with most acids and form salts, some of which exhibit remarkable differences occasioned by the oxides; thus, muriatic acid with the protoxide forms protomuriate of mercury, commonly called *calomel*, an insoluble salt; with the deutoxide it forms deutomuriate of mercury, commonly called *corrosive sublimate*, a soluble salt.

The proportions of metal and oxygen in the two oxides may be found by precipitating a known weight of mercury reduced by solution to either of the oxides, then drying and weighing the oxides, when the increase of weight by the addition of oxygen may be observed. This method is less accurate with regard to mercury than to other metals, on account of the very great weight of the atom, by which a small error in the gross weight of the oxide will be a great one as it respects the oxygen. This circumstance will partly account for the differences of authors on this subject.

One fact has been for some time known
which demonstrates the oxygen in the red
oxide to be double that in the black. Corrosive sublimate may be reduced to calomel
by adding to it as much mercury as the sublimate contains, and triturating the mixture
well, the oxygen of the red oxide (as well
as the acid) becomes equally divided amongst
the mercury and forms the black oxide, which
is a constituent of calomel. Hence it follows that if the oxygen in one oxide can be
ascertained, that of the other becomes known.
Or if we can find how much oxygen must be
added to the black oxide to change it to the
red, we shall know the oxygen in both.
Conformably with this last idea I have found
a very accurate and elegant method of ascertaining the oxygen required to convert the
black to the red oxide by treating protomuriate of mercury, mixed with water and a little muriatic acid, with oxymuriate of lime
in solution; this must be gradually added till
the protomuriate is dissolved, or rather converted to the deutomuriate. The quantity of
oxygen in a given solution of oxymuriate of
lime is most conveniently found by a solution
of green sulphate of iron, as will be shewn
under the oxides of that metal.

The oxides of mercury may be investigat-

ed by the nitrous gas produced during solution.
When mercury is dissolved without heat, as
mentioned above, no nitrous gas is liberated.
The solution has a strong nitrous smell and
requires a great quantity of oxymuriate of
lime to saturate both the oxide and the acid.
When heat is employed to accelerate the solution, nitrous gas is liberated. I dissolved
154 grains of mercury in nitric acid, 1.2
sp. gr., by the application of a gentle heat
from a lamp. About $\frac{1}{10}$ excess of acid remained in the solution; the nitrous gas obtained was 12 oz. measures $= 7.5$ grains, corresponding to 6.5 oxygen, which gives nearly
4 oxygen or 100 mercury. This would have
led me to suppose I had obtained the *black* oxide in solution; it was however entirely the
red, as it gave no precipitate by common salt,
and exhibited the red oxide by lime water;
but it required as much oxymuriate of lime as
contained 6.5 oxygen to saturate the *nitrous
gas* in the solution before any oxymuriatic acid
was liberated. It was clear therefore that
only $\frac{1}{2}$ of the nitrous gas was evolved, and
the other $\frac{1}{2}$ was retained in the solution,
though it had been exposed to boiling heat.

The following are the proportions assigned
by the several authors for the oxides of
mercury.

	Mercury.	Oxygen.	
		black.	red.
Bergman*	100 +	4 +	——
Lavoisier†	—— +	— +	7.75 to 8
Chenevix‡	—— +	12 +	18 5
Taboada‖	—— +	5.2 +	11
Fourcroy & Thenard (a)	—— +	4.16 +	8.21
Sefstrom (b)	—— +	3.99 +	7.99
My results give	—— +	4.2 +	8.4

Though the relative weights of oxygen and
mercury may be investigated as above, yet it
is from the weight of mercury and the acids
in the salts of mercury, some of which are of
a very definite character, such as the muriate and the deutomuriate, that the relative
weight of the atom of mercury is best investigated.   From these I first deduced the
weight of an atom of mercury to be 167
about 10 years ago, and subsequent experience has not induced me to change the number, though it probably may admit of some
correction.   If the atom of mercury be denoted by 167, that of the protoxide will be 174,

* Kirwan's Mineralogy.
† Annals of Philosophy, Vol. 3, p. 333.
‡ Philos. Trans. 1802.
‖ Jour. de Physique. 1805.
(a) Mem. d'Arcueil, Vol. 2. p. 168. 1809.
(b) Annals of Philosophy, Vol. 2. p. 48.

and that of the deutoxide 181; which makes 100 mercury take 4.2 and 8.4 oxygen for the oxides respectively, as in the above table.

5. Oxide of Palladium.

The discoverer of this metal, Dr. Wollaston, has given us its distinguishing chemical properties; but we are indebted to Berzelius and Vauquelin for the proportions of oxygen and sulphur which combine with the metal (Vid. Annal. de Chimie, 77 and 78.) Few chemists have had an opportunity of making experiments on this metal, owing to its great scarcity and the consequent high price of it (1 shilling per grain). It does not seem desireable that any but those skilled in the more delicate chemical manipulations should operate upon articles such as the present.

Berzelius treated the muriate of palladium with mercury, which abstracted the oxygen and left an amalgam of palladium and mercury; from the quantity of mercury dissolved he calculates that 100 palladium combine with 14.2 oxygen. This conclusion was corroborated by the circumstance that 100 palladium were found to take 28 of sulphur, or double the quantity of oxygen, which frequently happens with the metals.

Vauquelin precipitates the oxide of palladium from the muriate by potash; it appears of a red brown colour, and is probably a hydrate; when washed and dried in a moderate heat, it becomes black, it loses 20 per cent. by a red heat and becomes metallic. This would give 25 oxygen on 100 metal; but as he finds the sulphuret to be 100 metal with 24 or 30 sulphur, nearly agreeing with Berzelius, it is very probable that a moderate heat does not free the oxide from water, and that consequently a part of the 20 per cent. loss is water.

I dissolved 3 grains of palladium in a small excess of nitro-muriatic acid and obtained 240 grain measures of nitrous gas; the same quantity was obtained a second time, and to the solution (slightly acid) were added by degrees 200 measures of oxymuriatic acid gas; after agitation no smell was perceived, but by increasing the quantity of gas a permanent smell of oxymuriatic acid was produced, and when 200 more had been added the smell was sensible for some days in an open jar, a presumption that no higher oxide is to be obtained. Now 240 nitrous gas = .32 of a grain, corresponding to .28 of oxygen, and 200 oxymuriatic acid = .64 of a grain, cor-

responding to .15 oxygen; the sum of the
two portions of oxygen = .43, which must
have combined with 3 grains of palladium;
if .43 : 4 : : 7 : 50 nearly. Or 100 metal combine with 14 oxygen, as determined by Berzelius. I find the sulphuret to accord with
this determination; and by carefully saturating the excess of acid in the nitro-muriate
of palladium and then finding the quantity of
lime-water necessary to precipitate a certain
weight of palladium, as well as the quantity
of test muriatic acid necessary to dissolve the
precipitated oxide, I am confirmed in the
opinion that the atom of palladium must
weigh 50 nearly, and its oxide 57, which
there is every reason to believe is the protoxide.

6, 7, and 8. *Oxides of Rhodium, Iridium, and Osmium.*

Nothing certain has yet been determined
respecting the oxygenation of these very
rare metals.

9. *Oxides of Copper.*

There are two oxides of copper according
to the results of Proust, Chenevix, Berzelius and others, the proportions of which are

given nearly the same by all, and so as to leave no reasonable doubt concerning their accuracy.

1. *Protoxide.* This oxide is *orange*, and contains 12½ oxygen on 100 copper: it is obtained by precipitating a portion of copper from the solution of any cupreous salt, by means of iron, then mixing this copper with a rather greater portion of the deutoxide and triturating them well. This being done, the mixture is to be dissolved in muriatic acid, and the orange oxide may then be precipitated by an alkali.

2. *Deutoxide.* This oxide is *black*; it contains 25 oxygen on 100 copper: the *black* oxide is obtained by dissolving copper in nitric or sulphuric acid, then precipitating by lime-water or an alkali, and heating the dried precipitate red hot. It may also be obtained by exposing copper to a red heat for some time in common air or oxygen gas, removing the scales and exposing them in like manner, till at length the black oxide is formed.

By dissolving 112 grains of copper turnings in 1000 grain measures of 1.16 nitric acid, I obtained 48 oz. measures of nitrous gas, = 30 grains; by oxymuriate of lime I found 2 grains of nitrous gas in the solution, making in all 32 grains = 28 grains of oxygen.

If 28 : 112 :: 14 : 56, for the weight of an atom of copper; hence the protoxide = 63 and the deutoxide = 70.   These weights I adopted in 1806, and have not seen any reason to modify them since.

10. *Oxides of Iron.*

Two well known and well distinguished oxides of iron are now universally admitted; the one contains 28 oxygen on 100 iron, the other 42 on 100.

1. *Protoxide.*   This is always formed when iron is dissolved in dilute sulphuric or muriatic acid; it may be precipitated from these solutions by the pure alkalies or earths; it appears at first of a dark green, being then a hydrate or combined with water; on a filtre it soon becomes yellow at the surface by attracting oxygen; when dried in a heat of 200° or upwards it becomes black.   The quantity of oxygen in it is best ascertained from the hydrogen generated during the solution of the iron.   All the authorities I have found nearly concur in their results as under.

100 grains of iron dissolved in dilute sulphuric or muriatic acids yield hydrogen, according too

Cavendish   (1766)   155 cubic inches.
Priestley,   from 147 to 162
Lavoisier   163
Vandermonde, Berthollet, } max. 176
    and Monge   }
Vauquelin   160 to 179
Dr. Thomson   163
My own Experiments give   160
                                    ———
                        Mean 164=82 oxygen=
                                      27.9 grains.

By precipitating the oxide, and drying it, nearly the same result may be obtained, as 100 iron will yield 128 oxide. This oxide is magnetic.

2. *Intermediate or red oxide.* This oxide may be obtained in various ways. First by calcining the sulphate or nitrate of iron. Second by precipitation from old solutions of the salts of iron; the precipitate is *yellow* at first, being perhaps a hydrate; but when dried and heated it becomes brown-red. Third, by calcining iron or repeatedly exposing iron filings to a red heat, and trituration. Fourth, by treating a solution of the sulphate or other salt of the protoxide with oxmuriatic acid, or oxymuriate of lime till oxmuriatic acid is evolved; then precipitating the oxide which is thus converted into the red. Fifth, by agitating water containing the

green oxide recently precipitated, with oxygen gas. The red oxide is not sensibly magnetic.

The quantity of oxygen in the red oxide may be investigated in various ways, and it is generally allowed that they all concur in giving 42 on 100 iron. The one which I have used peculiarly, and prefer both for ease and accuracy, is to find the quantity of oxymuriatic acid gas necessary to saturate a given portion of the green sulphate. I take for instance 100 measures of 1.149 green sulphate, which I know to contain 8 grains of black oxide; this I find absorbs nearly 13 hundred measures of oxymuriatic acid gas before the acid smell is developed; the oxygen corresponding to this quantity of acid is known to be near 660 measures, $= .88$ grain. (See VOL. 1, p. 308.) Hence, if $8 : .88 :: 128 : 14$; or 128 black oxide acquire 14 or become 142 when converted into the red oxide. This fact being established, I find it very convenient to make use of the oxymuriate of lime instead of the acid gas, adopting the solution of green sulphate of iron as a test of the quantity of oxymuriatic acid in a given volume of any solution of oxymuriate of lime.

The quantity of oxygen in the red oxide of

iron may be inferred, but not so satisfactorily, from the nitrous gas obtained during the
solution of iron in nitric acid. In order to
obtain the most gas from a given quantity of
the materials, they should be so proportioned
as to produce saturation nearly. If an excess
of acid be used, it absorbs the nitrous gas in
part; and if an excess of iron, it is not all
dissolved. I took 50 grains of iron filings
and 600 measures of 1.15 nitric acid; these
were put together in a gas bottle and by the
assistance of a little heat a quantity of nitrous gas was obtained equal to 12 grains in
weight, allowing the sp. gr. of the gas to be
1.04 (air being 1); all the iron was dissolved
except a few atoms, and the solution was
slightly acid; the whole of the oxide was red
when precipitated by lime water. Now 50
grains of iron take 21 of oxygen to form the
red oxide, and these correspond to 24 of nitrous gas, which is just twice the quantity
obtained; one half of the gas generated then
remains in combination with the iron, even
when the constituents of the salt are proportioned so as to produce mutual saturation.
I was in expectation that the quantity of nitrous gas retained might be converted into
nitric acid by oxymuriate of lime, and hence
might be determined; but in this I was dis-

appointed. When oxymuriate of lime is
added to the liquid, a pungent gas is liberated, the nature of which I have not determined. Thinking it might in part be
owing to the iron, I transferred the acid
to soda, by decomposing the nitrate of iron
by the carbonate of soda; this nitrate of
soda however, when treated with oxymuriate of lime, exhibited the same pheno-
menon as the nitrate of iron. When an
acid is added the oxymuriatic acid itself is
given out. These results will require further
consideration. At present I am inclined to
think the pungent gas is one atom of nitrous
and one of oxygen or what I formerly considered as nitric acid. (See Vol. 1, plate 4,
fig. 27.)

Some authors have found as they conceive,
other oxides of iron, containing less or more
of oxygen than the above; thus Darso finds
by calcination from 15 to 56 oxygen on 100,
(Nicholson's Journ. Vol. 17); but there is
great reason to believe that uncertainties must
exist in his mode of experimenting sufficient
to account for the anomalies observed. This
author has suggested some doubt whether the
oxygenous gas naturally contained in water
has any effect on the salts with green oxide of
iron. I have ascertained that point by re-

peated experiments, and can assert that the oxygen in water immediately unites to the green oxide of iron to convert it into red, and that the green sulphate may be used as an accurate test of the quantity of oxygen in water. When pure green sulphate of iron is dropped into water and then the oxide precipitated by a gradual addition of lime water, it falls down yellow in proportion to the oxygen in the water, which may be increased 3 or 4 times by artificial impregnation. If the oxygen of the water be previously saturated with nitrous gas, then the oxide is wholly precipitated green.

Gay Lussac, in the 80th Vol. of the Annal. de Chimie, asserts that an oxide of iron containing 37.8 oxygen upon 100 iron is always obtained when iron is burned in oxygenous gas, and still more effectually when iron is oxydized by water or steam. If this oxide exist in the proportions stated, it must be a compound of 1 atom of the protoxide and 2 of the red oxide, which would give 37.3 oxygen on 100 of iron.

From the above facts and observations it is evident the atom of iron must be considered as weighing 25, (and not 50 as already given, Vol. 1, page 258); the protoxide is 32, and

the intermediate or red oxide is 2 atoms protoxide and 1 of oxygen = 71.

11. *Oxides of Nickel.*

1. *Protoxide.* It appears to be ascertained
from the experiments of Proust (Journ. de
Physiq.63—442), Richter (Nichols.Jour.12.),
Tupputi (An. de Chimie 78.), and Rolhoff
(An. of Philos. 3—335.), that the protoxide of
nickel consists of 100 metal and from 25 to 28
oxygen. My experiments on the solution of
nickel in nitric acid give me 14 grains nitrous gas, corresponding to 12 oxygen, in
the solution of 44 grains of nickel; this
gives 100 nickel to 27 oxygen, which I adopt
as agreeing with the mean of the beforementioned results. This oxide may be obtained
by precipitation from a solution of nitrate of
nickel; it is at first white, being then a hydrate; when dried in a moderate temperature
it becomes yellowish; after this, being heated
to a cherry red, it loses from 20 to 24 per cent.
of water and becomes of an ash grey colour:
this is the only oxide of nickel soluble in
acids, and must therefore be deemed the protoxide: hence we have 27 : 100 :: 7 : 26, nearly,

for the weight of an atom of nickel; and not 25 or 50, as estimated at page 258. VOL. 1.

Intermediate oxide. Thenard discovered a second oxide of nickel by passing oxymuriatic acid through a solution of nickel and then precipitating; it is a black powder; when treated with sulphuric or nitric acid it gives out gas, being the excess of oxygen above the protoxide; but with muriatic acid it gives oxymuriatic acid gas. Rolhoff was induced to believe, but I do not know upon what evidence, that this oxide contained $1\frac{1}{3}$ or $1\frac{1}{2}$ times the oxygen of the protoxide. By means of oxymuriate of lime I find the protoxide recently precipitated, takes half as much oxygen as it had previously, to form the black oxide; and that it cannot be formed, like the red oxide of iron, by agitation with water mixed with common air. The white oxide treated with oxymuriate of lime becomes almost instantly blue, growing darker till it gradually passes into brown, and finally black in about half an hour. It contains 40 oxygen on 100 nickel, and is most probably constituted of 1 atom of oxygen holding 2 of protoxide together, more especially as it is not found in combination with acids. The method I prefer to procure the black oxide is to precipitate a known weight of oxide by lime

water; then pouring off the clear liquid, I
put as much liquid oxymuriate of lime to the
moist hydrate as contains $\frac{1}{10}$ of the weight
of the oxide of oxygen, and stir frequently
for half an hour; the point of saturation is
found when more oxide put to the clear liquid
is not discoloured on one hand, and when
more oxymuriate of lime does not affect the
colour, but remains in the clear liquid on the
other hand.

12. *Oxides of Tin.*

There are two oxides of tin, which have
been carefully investigated by several chemists, and appear to be ascertained with great
precision. The protoxide is *grey*, and contains $13\frac{1}{2}$ oxygen on 100 tin; the *deutoxide*
is *white*, and contains 27 oxygen on 100 tin.

1. *Protoxide.* There are two methods of
obtaining the constitution of this oxide. The
first is by dissolving a certain weight of tin
filings in muriatic acid, precipitating by lime
water or carbonated alkalies and drying the
oxide in a moderate heat; this is liable to
some uncertainty; the precipitate being a *hydrate,* requires to be exposed to heat to expel
the water; but if the heat approaches to red,
the oxide takes fire and is converted into the

deutoxide. The second method is to dissolve tin in muriatic acid and carefully collect the hydrogen gas evolved; this was first done by Mr. Cavendish, with his usual accuracy, and published in 1766; he found 1 oz. of tin yield 202 oz. measures of hydrogen gas. I have frequently tried this experiment and always found a proportional quantity, or very nearly 200 measures for each grain of tin. This mode of investigation appears to me unexceptionable. Now 200 hydrogen unite to 100 oxygen, and 100 grain measures of oxygen = .134 grain in weight; hence if .134 oxy. : 1 tin :: 7 oxy. : 52 nearly for the weight of an atom of tin, on the presumption this is the protoxide.

2. *Deutoxide.* This may be obtained by heating tin till it takes fire, and the produce of the combustion is the oxide required; but to ascertain the proportions of tin and oxygen two other methods are preferable ; the one is to treat tin with nitric acid of the sp. gr. 1.2 to 1.4; a violent effervescence and great heat ensue and the tin is converted into a white powder. This being dried in 100° gives about 160 grains for 100 of tin. It consists of the deutoxide united to a little acid and water; these two may be driven off by a low red heat, and 127 grains of the deutoxide

remain in the state of a white powder. The other method is to treat a solution of the protoxide of tin with oxymuriate of lime till it is saturated; this will be found when 59 grains of the protoxide have acquired 7 grains of oxygen, or $113\frac{1}{2}$ have acquired $13\frac{1}{2}$ of oxygen, which corroborates the result by the 1st method. This oxide containing just twice as much oxygen as the former, may justly be considered as the deutoxide. No higher oxide of tin has been obtained.

The two oxides, though both white when precipitated, may be distinguished from their different appearances; the first is *curdy*, the second, *gelatinous*.

It may be proper to subjoin authorities for these oxides:

	Tin	Protoxide	Deutoxide
Cavendish, from the hydrogen	100	113.5	——
Proust (Journ. de Physique 59—341)	100	115	127$\frac{1}{4}$.128*
Gay Lussac (Annal. de Chimie 80—170)	100	113.5	127.2†
Berzelius (Annal. de Chim. 87—55)	100	113.6	127.2‡
My own, as above	100	113.4	127

* By nitric acid, the result of 3 experiments all agreeing for the deutoxide; the protoxide is by calculation and less certain. He afterwards adopts 13.6 from Berzelius. Journ. de Phys. Aug. 1814.

† The protoxide from hydrogen by solution; the deutoxide by transmitting steam over the metal at a red heat.

‡ The 2d. by oxydizing the sulphuret of tin by nitric acid; the 1st. by inference only, one half of the oxygen of the 2d.

13. *Oxides of Lead.*

There are three oxides of lead now generally recognized, the *yellow*, the *red*, and the *brown*, the proportion of oxygen in each of which has been investigated by several chemists whose results do not well accord with each other. I shall treat of them under the following names, namely the *protoxide*, the *intermediate oxides*, and the *deutoxide*, for reasons which will appear.

1. *Protoxide.* The yellow oxide of lead is the only one capable of forming salts with acids. Lavoisier found the oxygen of this oxide combined with 100 lead to be 4.47; Wenzel, 10; Proust, 9; Thomson, 10.5; Bucholz, 8; Berzelius, 7.7. This last accords best with my own experience; but it is chiefly from the other combinations of lead, that the weight of its atom as well as that of the protoxide are determined and confirmed, as lead forms several very definite compounds with acids, &c. The quantity of oxygen in the protoxide may be found by several methods, as under.

1st. By dissolving a given portion of the oxide in acetic acid, and precipitating the lead by another metal, as zinc; in this case the oxygen of the lead goes to the zinc which becomes dissolved, and from the loss of weight of the zinc and the proportion of oxygen in zinc oxide being previously known, and the weight of the precipitated lead being found, we have data for determining the oxide of lead. I took 200 measures of acetate of lead solution (1.142), which I knew contained 27 grains of oxide of lead; this being diluted with an equal volume of water, the lead was precipitated by a rod of zinc; in 6 hours an *arbor saturni* was formed which was collected and well dried; it weighed 21¾ grains, and the zinc rod had lost 7 grains: care must be taken in performing this experiment that all the lead be not precipitated, otherwise the oxide of zinc begins to fall, and the result is uncertain. In the residuary liquid I got 4 grains of sulphate of lead by sulphuric acid. Here then we have the oxygen of 21¾ lead transferred to 7 zinc; but if $7:21\frac{3}{4}::29:90$ nearly. Now it is known that 29 parts of zinc take 7 of oxygen, therefore 90 lead take 7 of oxygen, and the atom of lead=90, and the protoxide 97.

I formerly stated the atom of lead 95. VOL. 1, page 260.

2. By dissolving 180 grains of lead in nitric acid in a small thin capsule, and heating

it till the salt was quite dry, I got 288 grains
of salt, weighed in the capsule; 36 grains
of this salt yielded 24¼ yellow oxide by a low
red heat=22½ lead. This gives 90 lead to
7 oxygen.

3d. Again, 36 grains of the above salt,
dissolved in water, precipitated by ammonia,
and well washed on a filter, gave 23 + grains
of oxide separated from the filter, and this
had acquired 1 grain, making 24 + grains of
oxide from the 22½ lead as before; the residue of liquid gave no signs of lead by hydrosulphuret of ammonia. The same quantity
of salt precipitated by an excess of lime water gave only 22 grains of oxide; but hydrosulphuret of ammonia precipitated 2 + grains
of sulphuret of lead from the clear liquid.

II. *Intermediate oxide or oxides.* Minium
or red lead, &c. Minium is an article of
commerce used as a pigment and for various
other purposes. It is made by exposing the
yellow or protoxide of lead finely pulverized
to a low red heat in a current of air, and constantly stirring the oxide so as to expose fresh
particles to the air. In two days the yellow
oxide is converted into the red. Several authors observe that red lead usually contains
1, 2, or more grains per cent. of impurities
insoluble in nitric and acetic acids; the spe-

cimen I used however was so pure as not to leave more than $\frac{1}{3}$ of a grain per cent. of insoluble matter after being heated red and treated with dilute nitric acid.

Some of the most remarkable properties of red lead are, 1st. It is never obtained in combination with any acid; 2d. It yields oxygen gas when exposed to a bright red heat or when treated with concentrated sulphuric acid, and is in both cases reduced to the protoxide; 3d. When treated with dilute nitric acid it is dissolved in part, but constantly leaves an insoluble brown residuum, which is the deutoxide, as will be shewn; the weight of the deutoxide obtained is by my experiments 20 per cent. and the part in solution is found to be the protoxide; 4th. When treated with muriatic acid, muriate of lead is formed and oxymuriatic acid given out; 5th. When treated with dilute acetic acid or cold concentrated acetic acid, $\frac{1}{2}$ of the oxide is dissolved and the remainder is still red, its colour being rather improved; if concentrated acid be used and boiling heat applied, then $\frac{4}{5}$ of the whole oxide is dissolved and $\frac{1}{5}$ remains of brown oxide, the same as with nitric acid.

Some of the above facts are new, and may contribute to elucidate this most curious oxide, which scarcelyhas a parallel. Proust is the only author I know who has given a plausible conjecture concerning the peculiar nature of this oxide. He supposes it a compound of the yellow and brown oxides. This I believe is the fact; but it will be found I apprehend to be a compound of 1 atom of oxygen with 6 of the yellow oxide, as will appear from what follows.

Respecting the quantity of oxygen in the red oxide, Lavoisier finds 9 oxygen to 100 lead, Thomson 13.6, and Berzelius 11.55. This last is partly from experience and partly from a supposed analogy, that the successive oxides of the same metal contain oxygen as 1, $1\frac{1}{2}$ and 2 respectively; and having found (I believe) correctly, that the brown oxide contains just twice as much oxygen as the yellow, this ingenious and generally accurate author adopts the theoretic inference in this instance at least prematurely, and concludes the red oxide is the mean between the yellow and the brown. But we must appeal to experience.

It has already been stated that when red lead is exposed to heat, oxygen gas is given out, and it may be added, a small trace of water; and yellow oxide remains.

This experiment requires considerable
skill. If too great a heat is used, a part of
the lead is reduced or revived as it is termed;
if too little heat, then a part of the red lead
remains unaltered. In performing this experiment I use a small clean iron spoon to hold
the red lead, and cover it by another iron
spoon; the whole is then held by a pair of
tongs in a red fire till the spoon exhibits a
uniform moderate red, and some time after.
It is then withdrawn and cooled, and the
oxide weighed. The average loss of weight
is nearly 2 grains per cent. If only 1 grain
or less, a considerable portion of red oxide
remains mixed with the yellow; if 3 or more
grains, then the margin of the oxide exhibits particles of lead amounting to $\frac{1}{10}$, less
or more, of the original weight; this can be
easily separated from the oxide if necessary,
but it is apt to adhere to the iron; when red
oxide remains, it is so mixed with the yellow
as not easily to be separated, but its quantity
may be determined by nitric acid, which dissolves the yellow, and $\frac{4}{7}$ of the red, leaving
a residuum of brown oxide, from which the
quantity of red is inferred. Now if the loss
of weight of 100 red oxide be only 2 grains,
and a part of that be water, it is impossible

that 115.55 should lose 3.85 grains of oxygen,
according to Berzelius. Another experiment,
equally decisive of the question, is to determine the quantity of oxygenous gas to be obtained by heat or acids from a given weight
of red lead. In one experiment made with
great care, 500 grains of red oxide gave 6
grains of oxygenous gas by sulphuric acid;
in another, 200 yielded $2\frac{1}{2}$ grains. In order
to vary the mode of determining the quantity
of oxygen, into 210 measures of test green
sulphate of iron solution, $(1.156) = 16.8$ green
oxide, put 160 grains of minium; to this
was added dilute muriatic acid more than
sufficient for the minium: The oxymuriatic
acid from the oxygen of the minium was instantly seized by the oxide of the iron, the
whole of which was found by precipitation
to be changed from green to red and an excess of oxymuriatic acid appeared. Now
16.8 oxide would require 1.86 oxygen to become red, which it must have acquired from
160 of red lead; or 100 red lead yielded 1.2
oxygen, the same proportion as by sulphuric
acid. These experiments point out 1.2 oxygen in 100 red lead as the excess which converts the yellow to the red oxide. Were
any doubt to remain on the subject, the experiment with nitric acid and red oxide will

remove it.   If the red oxide contained a mean
of oxygen between the yellow and the brown,
when it is treated with nitric acid more than
50 per cent. of brown oxide would be obtained instead of 20, which is contrary to
all experience.   It must be observed that
Berzelius informs us he extracted the yellow
oxide, mechanically mixed (as he conceives)
with the red oxide, by digestion with dilute
acetic acid; but he does not inform us how
much per cent. his minium was reduced by
this operation.   From what is stated above,
it appears that about $\frac{1}{2}$ of the whole is thus
dissolved.   The remaining half would then
contain double the quantity of oxygen and
brown oxide per cent. that the original did.
Still these quantities are inadequate to explain
the phenomena.   Besides it cannot be admitted
that a *red* and a *yellow* powder can be intimately mixed in equal quantities and the
mixture not be distinguishable without difficulty from the *red* one, and be altogether different from the *yellow*.   We must then conclude that the minium of commerce (such as
I have used) is a true chemical compound.

   Grounding our reasonings upon the preceding facts, there are but two suppositions that
can be considered as plausible, respecting the
constitution of the red oxide.   It may be 1

atom of oxygen and 5 of yellow oxide, or 1 atom of oxygen and 6 of yellow oxide. The former would give 1.4 per cent. extra oxygen in 100 red oxide, and 21 brown oxide; the latter would give 1.2 per cent. extra oxygen and 18 brown oxide. I adopt the latter supposition; because it agrees with experiment in regard to oxygen, and gives the brown oxide a little *lower* than experiment, as may be expected on two accounts; first, the residue of brown oxide contains the insoluble dross of the red oxide (which was very small however, as stated above); and, second, unless a considerable excess of nitric acid be used, or long digestion, a small portion of the red oxide escapes decomposition. Another and still more important consideration, as to the question whether 5 or 6 atoms, is the equal division of the red oxide by the operation of cold acetic acid; it reduces the 1 oxygen and 6 yellow oxide to 1 and 3 atoms; whereas if we adopt the other, we must conclude it reduces the 1 and 5 to 1 and $2\frac{1}{2}$, a position that cannot well be reconciled to the atomic theory.

According to this conclusion then the red oxide of lead or minium of commerce is constituted of 1 atom of oxygen holding 6 atoms of yellow oxide together; or it is composed of 100 lead and 9.07 oxygen. When it is

digested in cold acetic acid the residuum constitutes another oxide consisting of 1 atom of
oxygen and 3 of yellow oxide, or 100 lead
and 10.4 oxygen, possessing the same colour
as the former, but distinguishable by its not
being acted on by cold acetic acid, and by its
containing twice as much brown oxide and
extra oxygen as minium. No doubt the other
intermediate oxides of 1 to 4 and 1 to 5 exist,
and are all alike red; but have not perhaps
any remarkable distinctions besides their
containing different proportions of oxygen
and brown oxide. Whether an oxide consisting of 1 oxygen and 2 yellow oxide exists, I
have not discovered; but that 1 oxygen and
1 yellow oxide are found united, appears
below.

III. *Deutoxide.* This is the flea-brown oxide mentioned above. It may also be obtained
by treating solutions of salts containing the
yellow oxide by oxymuriate of lime, in
which case the oxide is precipitated, leaving
the acid in the liquor, a proof that it is insoluble in acids. Its more remarkable properties are: 1st. like the red oxide, when heated
to a low red, or treated with sulphuric acid,
it yields oxygenous gas, and more copiously;
it is thus reduced to the yellow oxide: 2d.
with muriatic acid it yields oxymuriatic acid

in great plenty and muriate of lead: 3d. it
detonates when rubbed with sulphur in a
mortar.

The quantity of oxygen in the brown oxide is stated by Thomson at 25 oxygen to 100
lead, by Berzelius at 15.6 to 100. This last
is very nearly right by my experience, and
being just double of the oxygen in the protoxide, it warrants us in denominating it the
deutoxide. Berzelius finds 100 of the brown
oxide lose 6.5 by a red heat so as to reduce it
to the yellow; Dr. Thomson finds the loss 9
grains. This difference is easily accounted
for; it loses, I find, from 7 to 10 grains per
cent. according to the previous degree of
dryness; when exposed to a moist atmosphere
it attracts humidity; when dried in a temperature of 200° and exposed to red heat immediately after, it does not lose more than 6.5 or
7 per cent. This is corroborated too by the
oxygen expelled by sulphuric acid. From
100 grains of brown oxide and sulphuric acid
in a gas bottle, I obtained by the heat of a
lamp 8.3 oz. of oxygenous gas = 5.3 grains;
about 120 grains of grey sulphate of lead
were left in the bottle. The oxygen is rather
less than was expected; but it must be remembered that 100 grains of brown oxide,
obtained in the ordinary way, have the inso-

luble dross of 500 red oxide in them, which must have some influence in diminishing the production of oxygen.

Though the above might be deemed sufficient to demonstrate the proportion of oxygen in the brown oxide, I was desirous to corroborate the results by oxymuriate of lime. I found repeatedly that 100 grain measures of acetate of lead $(1.142) = 13.8$ yellow oxide, required 400 measures of oxymuriate of lime $= 1$ oxygen, to precipitate the whole of the oxide in a brown state.   Now if $13.8:1::$ $97:7$. Again, into 240 measures of test green sulphate of iron $(1.156) = 19$ oxide, were put 40 grains of brown oxide of lead, together with a sufficient quantity of muriatic acid to saturate the lead, and discharge the oxygen; after due agitation sulphate of lead was precipitated, and the whole of the oxide of iron was found, when precipitated, to be yellow.   But 19 grains oxide of iron require $2 +$ of oxygen to become yellow; hence the 40 grains brown oxide of lead must have furnished $2 +$ grains of oxygen to form oxymuriatic acid, which transferred it to the oxide of iron.   If $40:2 + :: 100:5 +$ oxygen, for the excess or second dose of oxygen in 100 brown oxide, such as is obtained by nitric acid along

with its impurities; which agrees with the
results obtained by the other methods.

14. *Oxide of zinc.*

When zinc is exposed to a strong heat it
burns with a brilliant white flame, and
a white powder sublimes, which is the oxide
of the metal. When dilute sulphuric acid is
poured on granulated zinc, hydrogen gas is
produced in great abundance and purity; the
metal is oxidized at the expence of the water
and dissolved in the acid, the oxide may be
precipitated by an alkali; it is white both
when precipitated and dried, and when heated
does not differ from that obtained by combustion. By a violent heat it runs into glass.

The quantity of oxygen in zinc oxide is, I
think, best estimated by the hydrogen gas
produced during the solution; it may also be
obtained by direct combustion, or by solution
in nitric acid and calcination. Dr. Thomson
determines the oxygen by comparison of the
weights of real sulphuric acid and metallic
zinc in a solution of sulphate of zinc, along
with the consideration that the proportion of
sulphuric acid and oxygen in the metallic
sulphates is known; Mr. Cavendish obtained
356 oz. measures of hydrogen from 1 oz. of

zinc by solution. I dissolved 49 grains of zinc in dilute sulphuric acid and obtained hydrogen, after the rate of 363 grain measures for 1 grain of zinc = 182 measures of oxygen = .24 grain of oxygen.

The following are the principal authorities for the quantity of oxygen in zinc oxide, in the order of time.

		Zinc.	Oxygen.
1766.	Cavendish	100	+ 23.3
1785.	Lavoisier	——	+ 19.6
1790—1800.	Wenzel and Proust ...	——	+ 25
1801.	Desorme and Clement	——	+ 21.7
	Davy	——	+ 21.95
	Berzelius	——	+ 24.4
	Gay Lussac	——	+ 24.4
	Thomson	——	+ 24.42
	My own	——	+ 24

Now if 24 oxy.: 100 zinc :: 7 oxy.: 29 zinc, nearly, which is therefore the weight of an atom of this metal, on the supposition that the oxide is 1 oxygen and 1 metal; and the atom of oxide = 36.

I formerly estimated the atom of zinc at 56 (VOL. 1, page 260). This was occasioned by taking the above as the *deutoxide* instead of the *protoxide*. By violently heating the oxide of zinc in a close vessel, Desorme and Clement reduced the oxygen nearly one half, so as to afford a presumption that an oxide

with half the oxygen of the common one sub-
sisted. Since that time some observations of
Berzelius seem to shew that a sub-oxide of
zinc exists. It does not appear however, that
such oxide is ever found in combination with
acids; and, granting the accuracy of the observations, it is rather to be presumed to be
the semi-oxide, or 1 atom of oxygen and 2 of
metal, than the protoxide. No higher oxidation of zinc than the above has yet been
obtained, and probably does not exist.

15. *Oxides of potassium.*

Since writing the articles "potassium and
sodium," in the former volume, a very important essay relating chiefly to these subjects has
been written by Gay Lussac and Thenard (a
copy of which they were so good as to send
me), entitled "Recherches Physico-chimi-
ques, &c." in 2 Vol.—Many of the most interesting experiments of Davy have been repeated on a larger scale, and a great number
of original ones added; these ingenious authors endeavour to sum up the evidences for
and against the two hypotheses concerning
potassium and sodium, namely, as to their being metals or hydrurets, and upon the whole
incline to the former, allowing however, that

the facts afford great plausibility to both. One thing they seem to have discovered and established, that the new bodies or metals admit of various degrees of oxidation, and of course these products have a claim to be classed amongst oxides in general though the nature of their bases may still be an object of dispute.

They find three oxides of potassium; the lowest degree is obtained by exposing potassium to atmospheric air in a small bottle, with a common cork; a gradual oxidation takes place; a blueish grey brittle product is obtained; there does not appear however, to be any proper limit to this oxidation besides that which they admit as characterizing the second degree or potash, which degree of oxidation may always be immediately obtained by placing potassium in contact with water. This I think should be called the protoxide and considered as 1 atom of potassium, and 1 of oxygen; before this point it is potassium and pot-ash mixed or perhaps combined.

Besides these there is another obtained by burning potassium in oxygen gas at an elevated temperature; this oxide is yellow, fusible by heat, and crystallizes in lamina on cooling; it contains three times as much oxygen as potash; put into water it is suddenly decomposed,

giving out $\frac{2}{3}$ of the oxygen in gas and becoming potash. Very probably an oxide containing twice as much oxygen as potash might be
formed with some mark of discrimination,
by uniting 18 parts potassium with 56 of yellow oxide, but this has not yet been done.

According to these conclusions the weights
of the oxides of potassium may be stated as
under.—Potassium 35, protoxide or potash
42, deutoxide (supposed to exist) 49, and
the yellow or tritoxide 56. Hence we have

	Potassium.	Oxygen.	
Protoxide (potash)	100 +	20	Gay Lussac & Thenard
		19	Davy
Deutoxide	100 +	40	(unknown)
Tritoxide	100 +	60	Gay Lussac & Thenard

One feels unwilling to admit of a *tritoxide*,
(and that perhaps the only one existing,)
when the deutoxide is unknown, were it not
upon good authority. The obscurity on this
subject may be removed by future experiments.

It may be proper to add that Gay Lussac
and Thenard concur with Davy in assigning
a much greater saturating power to potassium
and sodium than to the fused hydrates of potash and soda of equal weights. From the table, Recherches, Tom. 2, p. 214, it may be
deduced that 35 potassium require as much

sulphuric acid to saturate them as 50 or more
of the hydrate of potash; and that 21 sodium
are equivalent to 36 or 37 hydrate of sodium.
If these results are accurate, the weights of
potassium and sodium, considered as hydrurets, cannot be as we have deduced them at
pages 486 and 503, Vol. 1, namely, 43
and 29 respectively, but 35 and 21, as at
page 262.

16. *Oxides of sodium.*

Gay Lussac and Thenard find a suboxide
of sodium in the same way as that of potassium, and it is probably a compound of soda
and sodium : the remarkable oxidation which
produces soda is, I should imagine, the protoxide or one atom to one, as obtained by
placing sodium in contact with water. A
higher oxide is obtained as with potassium,
by burning sodium in oxygen gas with a vivid heat. It resembles the yellow oxide of
potassium in its appearance and properties.
The degree of oxidation varies in the differ-
ent experiments from $1\frac{1}{4}$ to $1\frac{3}{4}$ times the
oxygen of soda. It is probably a combina-
tion of the protoxide and deutoxide. Hence
the oxides of sodium may be as under; reck-
oning the atom of sodium 21, and soda 28.

	Sodium.	Oxygen.
Protoxide (Soda)	100 +	33¼
Intermediate oxide	100 +	50

17. *Oxide of bismuth.*

Only one oxide of bismuth is known, and the proportion of its parts has been gradually approximated by Bergman, Lavoisier, Klaproth, Proust, and others. Berzelius mentions a purple oxide obtained by exposing bismuth to the action of the atmosphere; but as no experiments have been made upon it, we cannot adopt it at present. According to Klaproth and Proust, 100 bismuth unite with 12 oxygen; but by the more recent experiments of Mr. J. Davy and Lagerhjelm 100 bismuth take 11.1 or 11.3 oxygen. If we adopt this last, which is doubtless near the truth; we shall have 11.3 : 100 :: 7 : 62 for the weight of the atom of bismuth, on the supposition that the compound is the protoxide or 1 atom of metal to 1 of oxygen. My former weight of bismuth was 68 (page 263), which is clearly too high.

Bismuth is best oxidized by nitric acid. Part of the oxide combines with the acid and part precipitates in the state of a white powder; if the whole be gradually heated, the acid is driven off, and at a low red the oxide remains pure; it is fused into glass and of a red or yellow colour, according to the heat employed. Bismuth may also be oxidized by heat in open vessels; yellow fumes arise which may be condensed and are found to be the oxide.

18. *Oxides of antimony.*

Considerable difference of opinions exists with regard to the oxides of antimony. Proust finds two oxides which he determines to consist, the first, of 100 metal + 22 or 23 oxygen; the second of 100 metal + 30 oxygen. Thenard finds 6 oxides: J. Davy two oxides, namely, 100 metal + 17.7 oxygen, and 100 + 30 oxygen. Berzelius infers from his experiments that there are 4 oxides of antimony, the first containing 4.65 oxygen, the second 18.6, the third 27.9, and the fourth 37.2 of oxygen on 100 metal. He admits however that the oxide obtained by boiling nitric acid on antimony and expelling the superfluous acid by a low red heat, consists of 100 metal + 29 to 31 oxygen, as determined by Proust and others. This is certainly the most definite of the oxides, next to that which is obtained from the solution of antimony in

muriatic acid. This last may be had by
pouring water into a solution of muriate of
antimony; a white powder precipitates, which
is the oxide with a little muriatic acid; the
acid may be abstracted by boiling the precipitate in a solution of carbonate of potash.
This oxide is a grey powder, and fusible at
a low red heat. It enters exclusively into various well known compounds, as the *golden*
sulphur of antimony, antimoniated tartrate of
potash, &c. Its constitution, according to
Proust, is 100 metal + 23 oxygen; but J. Davy finds only 17.7 oxygen, and Berzelius 18.6.
As this oxide possesses the most distinct fea-
tures, and besides is the most important, it is
desirable its constitution should be ascertained
without doubt. From several experiments I
made on the precipitation of antimony by
zinc, I conclude the oxide contains about 18
oxygen on 100 metal. I took the common
muriate of antimony with excess of acid, and
immersed a rod of zinc into it, covering the
whole with a graduated bell glass. Hydrogen
gas was produced by the excess of acid, and
its quantity was ascertained; the antimony
was in due time precipitated, and when the
operation ceased, the loss of zinc and the
weight of antimony were found. For instance, to 50 measures of 1.69 mur. ant. 60

water were added, no precipitation was observed; a zinc rod was put in and the whole covered by a bell glass, over water; in a few hours the operation had ceased, and there appeared 3480 grain measures of hydrogen gas generated; the dried antimony weighed 25½ grains, and the zinc had lost 29 grains. Now 3480 hydrogen require 1740 of oxygen =2.3 grains in weight. But 29 zinc require 7 oxygen; therefore the zinc must have got 4.7 oxygen from the antimony; that is, 25.5 antimony were found united to 4.7 oxygen; this gives 100 antimony + 18.4 oxygen. I conclude then that the error is with Proust; and this appears to be confirmed by the consideration that Proust himself obtains only 86 oxide of antimony from 100 sulphuret, which he allows to contain 74 antimony; now if 74: 12 :: 100 : 17 nearly. I am therefore inclined to adopt 18 for the oxygen which combines with 100 antimony to form the grey oxide. Whether this is the protoxide or deutoxide may be disputed; and the facts known concerning the other oxide or oxides will scarcely determine the case: but the proportions of the muriate and sulphuret of antimony accord much better with the former supposition. Now if 18:100 :7: 39, for the weight of the atom of antimony; I prefer the weight 40, deduced from the sulphuret, as announced in VOL. 1, page 264.

The oxide which contains 30 on 100 must be 2 atoms of the deutoxide and 1 of the protoxide united. What Berzelius calls the white oxide or antimonious acid, may be 1 atom of each oxide united, containing 27 oxygen on the 100. The oxide supposed to contain 36 or 37 oxygen on 100, and which must be considered as the deutoxide, has not been proved to exist separately. My efforts to procure it have failed as well as those before mine: by treating muriate of antimony with oxymuriate of lime I have obtained oxides of 30 on the 100, but never much higher. Whenever a greater proportion of oxymuriate of lime is added, the smell of the gas becomes permanent.

Antimony exposed to a red heat in a current of common air or oxygenous gas takes fire, and white fumes arise formerly called *flowers of antimony*; this oxide contains 27 or 30 oxygen on 100 metal.

Antimony thrown into red hot nitre is oxidized rapidly; the remaining powder, washed in water, is found to be a compound of oxide of antimony and potash. Berzelius calls the oxide the antimonic acid, and the salt the *antimoniate of potash.* It consists, according

to his experience, of 100 acid and 26.5 potash.   A similar salt formed between the antimonious acid and potash is constituted of 100
acid and 30.5 potash.

19. *Oxide of tellurium.*

We are chiefly indebted to Berzelius for
the proportions in which tellurium combines.
He finds 100 tellurium unite to 24.8 oxygen.
Also that 201.5 tellurate of lead gave 157
sulphate of lead.   This last contains 116 oxide of lead, which must therefore have combined with 85.5 of the oxide of tellurium.
Hence 97 oxide of lead would combine with
71.5 oxide of tellurium $= 57\frac{1}{2}$ tellurium $+ 14$
oxygen.   Whether this oxide of tellurium is
the protoxide or deutoxide, is somewhat uncertain.   The atom of tellurium will weigh
$57\frac{1}{2}$ in the latter case, but only 28 or 29 in the
former.   The analogy of the oxide to acids
favours the notion of a deutoxide; but the
facility with which the tellurium is volatilized
by hydrogen is in favour of the lighter atom.
The oxide is a white powder; it is produced
by dissolving the metal in nitro-muriatic acid
and precipitating by an alkali.

20. *Oxides of arsenic.*

There are two distinct combinations of arsenic and oxygen; the one has been long
known as an article of commerce under the
name of arsenic. It is a white, brittle, glassy
substance, obtained during the extraction of
certain metals from their ores. Its specific
gravity is about 3.7. According to Klaproth
boiling water dissolves from 7 to 8 per cent.
of the oxide of arsenic; but on cooling it
retains only about 3 per cent.; and this I find
is gradually deposited on the sides of the vessel till it is reduced to 2 per cent. or less in
cold weather, and by some months standing.
Water of 60° degrees or under dissolves no
more than ¼ per cent. of the oxide. At the
temperature of about 400° the oxide sublimes.
This oxide combines with the alkalies, earths,
and metallic oxides somewhat as the acids do,
but does not neutralize them, and in other
respects it is destitute of acid properties; as
for instance, it does not affect the colour tests.
It is extremely poisonous.

The other oxide is obtained by treating
either the white oxide or pure metallic arsenic
with nitric acid and heat. One hundred
grains of white oxide require two or three
times their weight of nitric acid, of 1.3, to

oxidize them. The new oxide is produced
in a liquid form; from which the excess of
nitric acid may be driven by a low red heat,
and the oxide is obtained pure in the form of
a white opake glass, which soon becomes liquid by attracting moisture from the atmosphere. This oxide, discovered by Scheele,
has all the properties of acids in general, and
is therefore denominated arsenic acid. When
just fluid by attracting moisture it has the
sp. gravity 1.65 nearly. It is represented as
equally poisonous with the white oxide.

The proportions of the elements in these
two oxides have been investigated with considerable success. Proust finds the white
oxide constituted of 100 metal and 33 or 34
oxygen, and the second of 100 metal with
53 or 54 oxygen: with these results those of
Rose and Bucholz nearly agree. Thenard
finds 100 + 34.6 for the white oxide, and 100
+ 56.25 for the acid: and Thomson 100 + 52.4
for the acid. Berzelius however, infers from
his recent experiments that the oxide consists
of 100 metal + 43.6 oxygen, and the acid of
100 + 71.3; these last results I have little
doubt are incorrect from my own experience.

It appears that when arsenic is oxidized by
nitric acid, 100 parts yield from 152 to 156
of acid, dried in a low red heat. The differ-

ences may in part be owing to the metal being partly oxidized at the commencement of the operation. On this account I should suppose 55 or 56 to be the proper quantity of oxygen united to 100 metal to form the acid. Proust and Thenard both found that 100 white oxide, when converted into acid by nitric acid, gave 115 or 116. I have found the same. Now if $116:100::156:134$; hence the white oxide of arsenic must contain 100 metal to 34 oxygen, if the data be correct; or the metal and oxygen are as 3 to 1 nearly. It is highly improbable that any inferior oxide subsists, as no traces of such have been found, if we disallow a conjecture of Berzelius on the subject. The white oxide of arsenic must then be considered as the *protoxide,* and the atom of arsenic must weigh 21 nearly, and that of the protoxide 28.

It is plain the other is not the *deutoxide,* as it does not contain twice the oxygen of the protoxide; but as the proportion of oxygen in it is to that of the protoxide, as $5:3$, it may be a compound of 2 atoms of deutoxide, and 1 of protoxide; that is, it may be the *superarseniate of arsenic,* if we consider the deutoxide as the acid, and the protoxide as the base. According to this view, the compound oxide, or *arsenic acid* of

Scheele, is constituted of two atoms of the deutoxide, weighing 70, and 1 atom of the protoxide weighing 28, together making 98, for the weight of an atom of arsenic acid,=63 arsenic + 35 oxygen: and 100 arsenic take 55.5 oxygen to form the acid, agreeably to the above recited experiments. Singular as this conclusion may appear, the truth of it is put beyond doubt, I think, by the following experiments.

I have repeatedly found that 28 parts of white oxide in solution are sufficient to throw down 24 parts of lime, from lime water, so as to produce 52 parts of arsenite of lime, and leave the water free from both elements. This confirms the notion of the atom of protoxide weighing 28.

If to 24 parts of lime dissolved in water we put 98 parts of dry arsenic acid, the compound remains in solution, and is perfectly neutral to the colour test, but so that the addition of a small quantity of either ingredient disturbs the neutrality. If to this solution 24 parts of lime dissolved in water be added, the compound remains a limpid solution, but is very limy to the test. If to this we put in like manner, 24 parts more of lime, the whole compound is thrown down, and yields, when dried, 170 parts of arseniate of lime,

the liquid being now free from both elements. Here we see first, two atoms of
the deutoxide, neutralized by two atoms of
base, namely, 1 of arsenic oxide, and 1 of
lime; but (second), when one atom more of
lime is added, an union of 2 deutoxide, and
3 of base is effected, which of course is an
alkaline salt; when (third) more of lime is added, the 2 deutoxide and the 1 protoxide each
attach 1 of lime, and form a still more alkaline salt, which being insoluble, is wholly
thrown down, most probably in a compound
state of 98 parts arsenic acid, combined with
72 parts lime.

In like manner, I find that 42 parts of potash, 28 of soda, and 12 of ammonia, seve-
rally neutralize 98 parts of arsenic acid.

1st.   24 lime + 32.7 arsenic acid = insoluble arseniate
2d.   ——   + 49   ——— —— = soluble arseniate
3d.   ——   + 98   ——— —— = neutral arseniate

It is a remarkable fact, that when neutral
arseniate of potash and nitrate of lead are
mixed together to mutual saturation, the precipitate is found to consist chiefly of arsenic
acid and oxide of lead, in proportion of 1
of acid to two of oxide, (that is, 98:194,
or 100:198); which does not differ much from
the determination of Berzelius.

I find however, only one fourth of the nitric acid in the residuary liquid in a free state; which leads me to suspect that the precipitate is a compound of subnitrate and arseniate of lead, in which the arsenic acid and lead are in due proportion, or 98 acid, to 97 oxide. This consideration may be properly resumed hereafter.

Hence we conclude, the atom of arsenic weighs 21 (and not 42, as at page 264, Vol. 1), that of the protoxide or common white arsenic, 28; and that of arsenic acid = 98, being a compound of 2 atoms of deutoxide, and 1 of protoxide. Or,

100 Arsenic + 33.3 oxygen = 133.3 protoxide
———   + 55.5 ——— = 155.5 arsenic acid

21. *Oxides of cobalt.*

There are at least two oxides of cobalt, the one blue, the other black. Authors differ as to the proportions of the elements. Proust states the blue oxide to consist of 100 metal, and 19 or 20 oxygen, and the black of 25 or 26 oxygen. Klaproth finds in the blue, 100 metal and 18 oxygen. But Rolhoff according to Berzelius, finds 100 metal and 27.3 oxygen in the blue oxide, and 40.9 in the black. I have taken some pains to investi-

gate these oxides, and have been able to satisfy myself in a good degree, respecting
their constitution. The blue or protoxide
consists of 100 metal and 19 oxygen, and
the black oxide of 100 metal, and 25 or 26,
very nearly as Proust determined.

Protoxide. By repeated trials I have found,
that if 37 parts of metallic cobalt be treated
with the due quantity of nitro-muriatic acid,
and a heat of 150°, a rapid solution takes
place,; and a disengagement of pure nitrous
gas; this being carefully collected, it will be
found to weigh 8 grains, and of course corresponds to 7 grains of oxygen; hence 37 cobalt, unite to 7 oxygen, to form 44 of the
blue oxide; and as this is the only oxide that
combines with acids, it must be considered
as the most simple or protoxide, being 1 atom
of metal (37), and 1 of oxygen (7). The
estimation of the atom of cobalt at 50 or 60,
(page 265), must therefore be corrected.

Compound oxides. When the blue oxide of
cobalt is precipitated from a solution, by an
alkali or lime water, and oxymuriate of lime
is gradually dropped in, the precipitate changes colour rapidly; it passes from blue to
green and olive, thence to a dark bottle green,
and finally becomes black; oxygen gas is
given out copiously when an excess of oxy-

muriate of lime is used. I find the additional oxygen requisite to convert the blue to the black oxide is what Proust states it, namely, $\frac{1}{3}$ of that necessary to form the blue; hence it must be considered as a compound of 1 atom of oxygen and 3 of the protoxide. Probably the other coloured oxides are 1 to 4, 1 to 5, &c. The protoxide is blue when precipitated, but it is supposed to contain water, or to be a hydrate; as it is dark grey when heated. The blue oxide in a short time after precipitation being still under water, changes to a yellowish or dead-leaf colour; which also appears to be a hydrate of the protoxide, as it dissolves in acids without giving out gas; and yields the blue oxide by an alkali. According to Proust, this hydrate contains 20 or 21 per cent. water. If we suppose the blue to be 1 atom oxide, and 1 water, the yellow hydrate may be 1 water and 2 of the proto-hydrate; or 88 oxide, and 24 water, which will be nearly 21 per cent. water.

The black oxide gives out oxygen gas by a red heat, and is reduced to the grey oxide: it forms oxymuriatic acid, with muriatic acid, and the protoxide remains in solution.

(See Tassaert.—An. de Chimie 28; Thenard, 42; and Proust, 60.)

22. *Oxides of manganese.*

One of the oxides of manganese being a natural production, and sometimes of great purity, and the metal not being obtainable without skill and labour, it may be most convenient to adopt the inverse method in our investigations; that is, to trace out the atom of metal from its oxides.

Native oxides of manganese. Of late, I have met with excellent specimens of this oxide; they are in masses of a grey, crystalline appearance, sp. gr. 4, easily pulverizable into a greasy, shining, dark grey powder. They are nearly pure oxide; but the more common sort is blacker, and contains less or more of siliceous earth. Some specimens are very harsh, require an iron mortar to pulverize them, and contain 50 or upwards per cent. of siliceous earth. Of the common sort when pulverized, the black inclining to blue, is generally preferable to the black inclining to brown. I have not observed any earthy carbonates mixed with the oxide of manganese. Amongst various specimens I obtained the following analyses.

	Oxide.	sand and insoluble matter.
1. Grey, crystallized oxide............	100	——
2. Pulverized black oxide, from a bleacher, reputed good	80	20
3. Another specimen, in the lump	77	23
4. A light brown oxide	47	53
5. A sparry oxide, abounding with flint; black brown when pulverized	27	73

Some of the chemical characters of the native oxide of manganese are, its giving oxygen gas by a red heat, its insolubility in nitric and sulphuric acids, and its solubility in muriatic acid, but with the accompanying circumstance of disengaging oxymuriatic acid.

All these facts shew that it is of the higher order of oxides, or analogous to the brown and red oxides of lead.—The muriatic acid solution abovementioned, contains an oxide of an inferior degree, which is soluble in all acids, and which is the only oxide of manganese that appears to be soluble in acids. If this be considered, (as it may with the greatest probability), the protoxide, then it will appear from what follows, that the common native manganese is the deutoxide, and that there is an intermediate one, which contains a mean quantity of oxygen.

Protoxide. This may be obtained in solution with muriatic acid as above, from the native oxide. Or the black oxide may be mixed with sulphuric acid into a paste, and heated in an iron spoon to redness; the mass being lixiviated, a solution of the protoxide in sulphuric acid is obtained, generally with a slight excess of the acid; in this process heat and the presence of sulphuric acid, expels the redundant oxygen of the black oxide, and reduce it to the protoxide, which hence becomes soluble. If in either of these solutions any oxide of iron be present, whether from the manganese, or acquired during the manipulation, it is easily discovered and separated, as I have frequently found. Into any solution containing a mixture of the oxides of manganese, the green oxide of iron, and the red oxide of iron, let lime water be gradually poured; the red oxide of iron will be first precipitated, next the green oxide, and lastly the oxide of manganese, which may hence be separated from each other. Iron may also be discovered and separated by carbonate of potash, which must be dropped into the solution as long as any coloured precipitate appears; as soon as it has subsided, the snow-white carbonate of manganese succeeds. This white carbonate may be very conveniently used for obtaining solutions of pure manganese in any of the acids.

When a solution of pure manganese is treated with lime water, or ammonia, a light buff oxide, not much differing in appearance from the yellow oxide of iron, is obtained. This oxide is soluble in all acids, when recently precipitated; but, such is its avidity for oxygen, with moderate agitation of the liquid it acquires oxygen and becomes brown, when it ceases to be totally soluble; if dried in the air quickly, it becomes brown and obtains considerable oxygen. The buff oxide recently precipitated, is probably a hydrate; for, when the white carbonate of manganese is heated gradually to red, the water and the acid are both expelled, and a grey powder remains; this is quite black on the surface of the mass, if exposed to the air during the process. Probably this grey powder is the pure protoxide; it is soluble in acids, except the black powder at the surface; perhaps but for the oxygen of the air, the protoxide would be nearly white.

From its combinations with sulphuric and carbonic acids, I find the weight of an atom of the protoxide to be 32, or the same as that of iron. Dr. John, a German chemist, who seems to have investigated these salts with

more attention than any other person, has deduced nearly the same results. (Annals of
Philos. 2—172). He finds $33\frac{2}{3}$ sulphuric acid
+ 31 oxide, and 34.2 carbonic acid + 55.8
oxide; that is, when reduced to compare with
my results, 34 sulphuric acid + 31.3 oxide,
and 19.4 carbonic acid + 32 oxide.   This near
agreement may be considered as a confirmation of the accuracy of both.   Dr. John finds,
as I have done, three distinct oxides of manganese, the greyish green, the brown, and
the black.   The first of these is the only one
that combines with acids; but we differ materially as to the quantity of oxygen in each.
He found manganese decompose water at the
ordinary temperature; by oxidizing the metal this way, 100 metal acquired 15 oxygen
to constitute the protoxide; according to this,
28 metal + 4 oxygen would make 32 protoxide; but this conclusion would be so contrary to all analogy, that it cannot be admitted as satisfactory.   The probability is, that
the manganese must have contained a little
oxygen at the commencement of the experiment.   The general analogy of manganese
to iron, lead, &c. requires that 32 protoxide
should contain 7 oxygen.   If this be allowed,
we have the atom of manganese = 25, and
not 40, (as at page 266, VOL. 1), the same as

that of iron: and this conclusion is corroborated by what follows.

2. *Intermediate or olive brown oxide.* This
may be formed by combining oxygen directly
with the buff or protoxide recently precipitated, and still remaining in the liquor; simple agitation in oxygenous gas or common air
for a few minutes, is all that is requisite. Or
it may be instantly formed by treating the
same moist protoxide with liquid oxymuriate
of lime. Or it may be had by exposing the
purest black oxide to a bright red heat for some
time, when it will lose 9 or 10 per cent. and
there will remain the olive brown oxide.

To find the proportion of oxygen absorbed,
I precipitated 3.2 grains of the protoxide by
lime water; the liquid containing the oxide
was put into a well stoppered bottle of oxygen
gas; on agitation the oxide changed colour
fast, from buff to brown; in a short time it
absorbed 260 grain measures of gas = .35 of
a grain in weight, and then ceased to absorb.
In another experiment, 3.2 grains of precipitated protoxide, took 100 measures of a
solution of oxymuriate of lime, containing
.35 per cent. of oxygen, (that is, 1.45 oxymuriatic acid). Hence as 32 take 3.5, 64
must take 7; which shews the brown oxide to

be a compound of 1 atom of oxygen, and 2 of the protoxide.

The characters of this oxide are, its olive brown colour, its insolubility in nitric and sulphuric acids, without heat or deoxidation, and its solubility in muriatic acid after the evolution of oxymuriatic acid. By long exposure to the air, it is gradually changed, in all probability into the black oxide.

3. *Deutoxide.* In order to determine the quantity of oxygen deducible from the purest native oxide of manganese, to convert it into protoxide, I have successfully adopted the two following methods. 1st. Let 39 or 40 grains of the oxide be mixed with 60 common salt; to this add 80 grains of water, and 120 grains weight of strong sulphuric acid, in a gas bottle. The heat must be gradually raised to boiling, and the oxymuriatic acid gas may be received in a quart of lime water. This will be found sufficient to convert 800 measures of test green sulphate of iron (1.156) into red; that is, it will produce 29 grains of oxymuriatic acid, which will cause 7 grains of oxygen, to unite to the green oxide of iron. Now 100 measures of 1.156 sulphate, according to some recent experiments of mine, contain 8 grains of green oxide, (I estimated the sp. gr. of test sulphate, heretofore

at 1.149); hence 800 contain 64 oxide, and
these require just 7 grains of oxygen to be
united to them, to form the red oxide, as has
been shewn, page 34. In the above experiment, the 39 grains of oxide, will be found
to vanish or be dissolved, if pure, and to
yield 32 grains of protoxide, making up with
the 7 grains of oxygen, the original weight.
Hence we have 39 grains of the oxide resolved into 32 protoxide, and 7 oxygen. If
then we allow 32 protoxide, to contain 7
oxygen, it appears that 39 grains of the native oxide, consists of 1 atom manganese
(25), and two atoms of oxygen (14); or
it is the deutoxide of the metal. 2d. A
more direct and expeditious method, of
transferring the oxygen from the manganese to the iron, is as follows: Let 39 grains
of pure grey shining oxide, be mixed with
800 of test green sulphate of iron; to this
mixture let 25 or 30 grain measures of strong
sulphuric acid be added: after stirring the
mixture for 5 minutes, the oxide of manganese will be completely dissolved, and, on
precipitating the oxide of iron gradually, by
lime water, it will be found to be wholly *yellow* or buff; shewing that 7 grains of oxygen
have been transferred from the oxide of manganese to that of iron.—If more green sul-

phate of iron be used, then the surplus of
the oxide will be thrown down green; the
order of precipitation being the yellow oxide
of iron, the green oxide of iron, and lastly,
the yellow or buff oxide of manganese, as
has been stated. This affords an easy and
elegant method of appreciating the different
oxides of manganese of commerce; and it
was in this mode, the valuations of the specimens in the above table were made.

The proportions of the three oxides are
then as under:

	Manganese		Oxygen		
Protoxide	100	+	28	—	buff; soluble in acids.
Intermediate oxide	——	+	42	—	brown; insoluble.
Deutoxide	——	+	56	—	black; insoluble.

It may be proper to subjoin the results of
others, who have investigated the oxides of
manganese. Bergman finds 3 oxides, containing 100 metal + 25, 35, and 66.6 oxygen;
Dr. John finds 3 oxides, containing 100 metal + 15, 25, and 40 oxygen: Berzelius finds
5 oxides, containing 100 metal + 7, 14, 28,
42, and 56 oxygen; and Davy finds 2 oxides,
containing 100 metal + 26.6, and 39.9 oxygen, respectively.

23. *Oxides of chromium.*

There appear to be at least two oxides of chromium, one or other of which is found in combination with the oxides of lead or iron, but hitherto so very sparingly that few chemists have had an opportunity of investigating the proportions of chrome and oxygen, in the oxides of chromium. The chief sources for information on this subject, are essays by Vauquelin, An. de Chimie, VOL. 25 and 70; by Tassaert, *ibid.* 31; by Mussin Puschin, *ibid.* 32; by Godon, *ibid.* 53; by Laugier *ibid.* 78, and by Berzelius, Annal. of Philosophy, 3.

The oxides of chromium, as might be supposed, are distinguished for the colours which they possess and impart to the compounds into which they enter. One of the oxides is green; it gives colour to the emerald. The other is yellow, dissolved in water, but deep red when crystallized, and possesses the characters of an acid; it unites with alkalies, earths, and metallic oxides; it was first found in Siberia, in combination with the oxide of lead, a salt now denominated *chromate* of lead, of a splendid yellow colour, inclining to orange or red. Since then, the chromate of iron, has been found in France, America, and Siberia, with a prospect of greater abundance.

In order to investigate the weight of the atom of chromic acid, it is necessary to attend to such of the chromates as have been carefully examined. The chromates of potash, barytes, lead, iron, and mercury, are those with which we are best acquainted.

Vauquelin has given us the components of the native chromate of lead by analysis, and those of the artificial chromate by synthesis; the results do not accord very nearly: for, according to the analysis corrected by the modern science,

Chromate of lead............ $= 62$ acid $+$ 97 oxide

By synth. chromate of lead $= 57\frac{1}{2}$ — $+$ 97 ——

Berzelius however, has more lately given us the results of his experience, both analytical, and synthetical; and he finds both to give chromate of lead nearly $= 44$ acid $+$ 97 oxide.

Chromate of barytes...... (Vauq.)$=47.$S acid$+68$ barytes

Ditto.................... (Berz.)$=44$ —— $+68$ ——

Native chromate of iron (Vauq.)$=45$ acid $+35\frac{1}{4}$ oxide

Ditto................. (Laugier)$=55$ —— $+35\frac{1}{4}$ ——

Having received a small portion of chromate of potash in solution, from a chemical friend (J. Sims), I endeavoured to satisfy myself, as far as my materials would go, as to the nature and proportions of the chromates. The solution was of the sp. gr. 1.061, and consequently in 100 measures contained nearly 6.7 grains of chromic acid and potash, &c.—The liquid was a beautiful yellow; it was alkaline by the colour test. By the usual tests, I had reason to believe, that the solution contained as under per cent.— namely,

2.2 gr.	chromic acid
2.	potash
.8	uncomb. potash
1.4	carb. potash
.3	sulphate of potash
6.7	

With this liquid neutralized by nitric acid, I formed the chromates of lead, barytes, iron, and mercury; and I am inclined to believe these salts are nearly constituted as under:

Neutral chromate of potash   46 acid $+$ 42 potash
———— of barytes 46 ——— $+$ 68 barytes
———— of lead   46 ——— $+$ 97 oxide
———— of iron   46 ——— $+$ 32 oxide (black)
———— of mercury 46 ——— $+$174 oxide (black)

According to these results, the atom of chromic acid weighs 46; it is made 44 by the results of Berzelius, and from 45 to 62 by those of Vauquelin; I would not be understood to place great confidence in the above results of mine, though I am persuaded they will be found good approximations.

Is the chromic acid the deutoxide, or the tritoxide of chromium?

The determination will evidently be affected by the question, how much oxygen must be abstracted from the chromic acid to reduce it to the green oxide. Vauquelin finds 46 acid to lose $6\frac{1}{2}$ oxygen, and Berzelius $10\frac{1}{2}$, when converted into green oxide by heat. From the former of these, one would infer chrome to be 32, the green or protoxide of chrome to be 39, and the acid or deutoxide 46: from the latter, chrome = 25, protoxide = 32 (unknown), the green oxide = 1 protoxide and 1 deutoxide united [= 71 = 50 chrome + 21 oxygen = (25 chrome + $10\frac{1}{2}$ oxygen)$\times$ 2= $35\frac{1}{2}$ $\times$ 2] the deutoxide = 39, and the tritoxide or chromic acid = 46. I have not had an opportunity to perform any experiment that appears to me decisive as to the accuracy of one or other of these views; but shall make a few remarks relative to them.

The green oxide being the most prominent compound next to the chromic acid, being commonly produced from it by any deoxidizing process, being the lowest oxide known, and combining with acids, is on these accounts entitled to the consideration of the protoxide; indeed there does not seem an instance where the protoxide of a metal is unknown, whilst the deutoxide and compound oxides are known. There is however, another oxide observed by Vauquelin and by Berzelius, which is obtained by heating the nitrate, or combination of nitric acid and the green oxide, to dryness and expelling the acid; this oxide is brown, and gives oxymuriatic acid when treated with muriatic acid; on this account it would seem to be intermediate between the green oxide and the chromic acid: it is probably a combination of the two, or the *chromate of chromium*. On the other view however, it must be considered as the deutoxide. What corroborates the notion of the green oxide being 39, is the fact which I have observed, of 46 parts of chromic acid combining with 64 of the green oxide of iron to form 110 of chromate of iron; in this combination the oxide of iron may be said to borrow 1 atom of oxygen from the chromic acid, and the compound may then be considered as the union of the green oxide of chrome, and the red oxide of iron. When this precipitate is subjected to the action of muriatic acid, a green solution is obtained containing the oxide of chrome, and red oxide of iron is precipitated, as Vauquelin has observed. To form the above chromate (or rather subchromate) of iron, let a given portion of neutral chromate of potash be treated with green sulphate of iron, and lime-water be added, sufficient to saturate the sulphuric acid, a brown red precipitate is obtained; more sulphate and lime water must be gradually added to the clear liquid till the precipitate become green, when the proportions will be found as above stated. This artificial compound seems a subchromate; whereas the native compound seems to be a chromate. That there is some uncertainty in decomposing a chromate by heat with a view to obtain the green oxide, I have reason to suspect from having decomposed $5\frac{1}{3}$ grains of chromate of mercury by a moderate red heat; this compound contained 1.1 chromic acid, and it yielded only .6 of green oxide, whereas it should have been .9 or .8 at least.

Upon the whole I think the evidence is in favour of the opinion that the atom of chrome is 32, the green or protoxide 39, and the deutoxide or chromic acid is 46.

24. *Oxides of uranium.*

There appear to be two oxides of uranium from the experiments of Klaproth, Bucholz, and Vauquelin; but the proportions of metal and oxygen have not been very nearly ascertained, from the great scarcity of the minerals containing this metal. (Vid. Bucholz, An. de Chimie, 56—142. Vauquelin, ibid. 68 —277; or Nicholson's Journ. 25—69). The oxides are obtained by precipitation from solutions of the minerals in the nitric or muriatic acid, the foreign substances being first separated.

The protoxide of uranium precipitates dark bottle green by caustic alkalies, and forms crystallizable salts with acids; the other, probably the deutoxide, precipitates orange yellow, and forms uncrystallizable salts with acids; in these respects the oxides bear a near resemblance to those of iron.

Bucholz estimates the yellow oxide at 100 metal + from 25 to 32 oxygen; as it yields oxymuriatic acid when treated with muriatic, it is most likely to be the deutoxide; now if

we take 28 for the oxygen combined with 100
metal, the protoxide must consist of 100 metal + 14 oxygen, or of 50 metal + 7 oxygen,
and the atom of uranium = 50. From his account of the sulphate and nitrate of uranium
the weight of the atom might be inferred to
be double of the above or 100. These different conclusions can only be elucidated by future experiments.

25. *Oxides of molybdenum.*

The latest and as it should seem most accurate experiments on the oxides of molybdenum were made by Bucholz. (Vid. Nicholson's Journal, 20, p. 121). There appear to
be 3 oxides or combinations of molybdenum
and oxygen, namely, the *brown*, the *blue*, and
the *white* or *yellow*. The two last have the
character of acids, and none of them seem to
form salts with acids, like oxides in general.
Bucholz ascertained the above gradation,
and that the white oxide or molybdic acid contains $\frac{1}{7}$ of its weight of oxygen; (which has
since been corroborated by Berzelius); he also
found that the blue was best formed by mixing,
triturating, and boiling in water 3 parts of
brown oxide, and 4 of white, or one of metal, and two of acid; and that it has acid

qualities as well as the white.　Bucholz also
found 3 parts of liquid ammonia of the sp. gr.
97 dissolve 1 of molybdic acid; now 3 parts
of ammonia = .186 real (VOL. 1, p. 422);
and 1:.186 :: 64: 12, the quantity of ammonia usually saturated by one atom of acid;
and Berzelius found 100 molybdic acid
saturate 155 oxide of lead, or 63 acid to 97
oxide.　The native sulphuret of molybdenum
(the state in which this metal is usually found)
was analyzed by Bucholz and found to consist
of 60 metal and 40 sulphur.

The molybdic acid may be obtained by
roasting the sulphuret in a crucible and stirring it frequently; the sulphur in great part
escapes in the form of sulphurous acid and
the metal becomes oxidated: carbonate of
soda in solution may be added to the residuum
as long as any effervescence is observed; mo-
lybdate of soda remains in solution and the
acid may be precipitated by nitric acid.　The
brown oxide is best obtained by heating mo-
lybdate of ammonia to red; the ammonia and
part of the oxygen are expelled, and the
brown oxide remains.

There are two views with which the preceding results may be reconciled; namely, 1st.
supposing the atom of molybdenum to weigh
21; and 2d, by supposing it to weigh 42 or

twice that number. In the first case the brown oxide will weigh 24½ (49) being supposed 2 atoms of metal and 1 of oxygen, the blue or protoxide will weigh 28, and the white oxide or molybdic acid will weigh 63, being a compound of the protoxide and deutoxide, molybdena or native sulphuret will then be as usual, the protosulphuret, consisting of 21 metal and 14 sulphur, or 60 metal and 40 sulphur. In the 2d. case the brown or protoxide will weigh 49, the blue or deutoxide 56, and the acid or tritoxide 63. The native sulphuret, molybdena, must in this view be the deutosulphuret, or 42 metal and 28 sulphur.

The former of these views exhibits the oxides somewhat complicated, but agrees well with the sulphuret; the latter shews the oxides in a more regular train, but does not appear so probable from the sulphuret; besides, the notion of a metallic tritoxide is rather singular, especially in a metal that is rarely if ever found in combination with oxygen. Upon the whole I prefer the former view; but it must be considered as problematical only. The atom of 60 (see page 267 VOL. 1) must doubtless be erroneous.

26. *Oxides of tungsten.*

From the experiments of **D'Elhuiarts, Bu-**
cholz* and **Berzelius†** it seems very probable
that the tungstic acid is composed of about
100 metal + 25 oxygen. It is a *yellow* powder of the sp. gr. 6.12, and is best obtained
from the native tungstate of lime (a scarce
mineral). One part tungstate of lime and
four of carbonate of potash are fused together,
dissolved in water, and then the tungstic acid
may be precipitated by nitric acid. There is
an inferior oxide that is black or dark brown;
Berzelius reduced the yellow oxide to a flea-
brown colour, by sending a current of hydrogen gas through it in a glass tube heated
red hot. 100 parts of this oxide burnt be-
107 yellow oxide. Hence 100 metal must
combine with about 16½ or 17 oxygen to form
this oxide, which is $\frac{2}{3}$ of that in the yellow
or tungstic acid.—Upon the whole it does
not seem improbable, considering the great
sp. gravity of this metal, that it forms three
oxides and that the acid or yellow oxide is

* An. of Philos. 6—198
† An. of Philos. 3—244

the 3d. Hence the atom of tungsten must
be 84, that of the protoxide 91, the deutoxide 98, and the tritoxide or tungstic acid 105.
The native tungstate of lime, if pure, according to this would be 81.4 acid +18.6 lime,
which is not far from Klaproth's analysis; he
having found 18.7 lime in one specimen; nor
from that of Berzelius, he having found
80.4 tungstic acid and 19.4 lime in 99.8
tungstate of lime.*

There is another view however, which
would accord with the experiments and perhaps will be found preferable in other respects; that is, to suppose the tungstic acid to
be composed of 1 atom deutoxide and 1
atom protoxide united; in this case the atom
of tungsten = 42, that of the protoxide = 49,
that of the deutoxide = 56, and the tungstic
acid = 105 as before.

27. Oxides of titanium.

Nothing certain is known respecting the
oxides of titanium. An observation of Richter, quoted by Berzelius (An. of Philos.
3—251), if it could be relied upon, furnishes
an important fact, namely, that a solution of

*An. of Philos. 8—237

muriate of titanium containing 84.4 oxide, gave 150 muriate of silver. Now 150 muriate of silver contain 28 acid; hence 28 acid must have combined with 84.4 oxide; but if 28 : 84.4 :: 22 : 66 nearly for the weight of an atom of the oxide. This would indicate 59 for an atom of the metal.

28. *Oxides of columbium.*

The white oxide or acid of columbium is found in combination with the oxides of iron and manganese in proportion nearly as 4 of the acid to 1 of the aggregate oxides. The two minerals, columbite and tantalite, though yielding these substances nearly in the same proportions, are found to differ remarkably in specific gravity, the former being about 5.9 and the latter about 7.9. Dr. Wollaston concludes however, from the agreement of the white oxides extracted, that they must be the same. The white oxide of columbium is insoluble in the mineral acids; it unites with potash by fusion, and may be precipitated by most acids. Some of the vegetable acids, the oxalic, the tartaric, and the citric dissolve the white oxide. When the alkaline solution of columbium previously neutralized by an acid is treated with infusion

of galls, an orange precipitate is produced
which is characteristic of columbium. Nothing certain has been determined respecting
the proportions of metal and oxygen; but
from the great proportion of the columbic
acid found with the oxides of iron and manganese, together with the great sp. gravity of
the compound, one may pretty clearly infer
the great weight of the atom of columbium.
Supposing the white oxide or acid to consist
of 1 atom metal + 3 oxygen and that the columbite is formed by 1 atom of acid to 1 of
oxide, we should have 128 acid + 32 oxide.
This would give 107 for the weight of an atom
of metal, and 128 for that of the tritoxide
or columbic acid; but it is unnecessary to
dwell upon such conjectures.

In a recent memoir of Messrs. Gahn,
Berzelius, and Eggertz (An. de Chimie,
Octo. 1816), it is maintained as probable that
there is only one oxide of columbium or tanta-
lum, and that 100 metal take 5.485 oxygen,
or 121 metal take 7 oxygen. If this be correct, the atom of columbium must be 121 and
the protoxide 128.

(See also An. de Chimie, 43—271; Philos.
Trans. 1802; Nichols. Journ. 2—129; ibid.
3—251; ibid. 25—23).

29. *Oxides of cerium.*

The mineral cerite is of the sp. gr. 4.53, and constituted of 50 or 60 per cent. of oxide of cerium, with silex, lime, and iron. This mi- being calcined and dissolved in nitro-muriatic acid, the solution is to be neutralized by caustic potash, and then treated with tar- trate of potash. The precipitate, well washed and afterwards calcined, is pure oxide of ce- rium. This oxide, which is white, when calcined in the open air becomes red and ac- quires more oxygen. These oxides, particu larly the white, are soluble in most acids; the red oxide with muriatic acid gives oxy- muriatic acid.

The experiments hitherto made on this sub- ject scarcely enable us to decide respecting the proportions of metal and oxygen, nor the relative weights of these oxides.

Both Vauquelin* and Hisinger† agree that the proto-carbonate of cerium, when expo- sed to a red heat, yields 57 or 58 oxide, which the former says is the red oxide, being

* An de Chimie, 54—28
† An. of Philos.—4-356

changed by the calcination. Hisinger finds
the percarbonate to consist of 36.2 acid and
63.8 oxide: also that the muriate of cerium
consists of 100 acid and 197.5 oxide; but
Vauquelin remarks that the sulphate, nitrate,
and muriate of cerium are always more or
less acid, however dried; and he found the
protoxalate of cerium to yield 45.6 red oxide
by calcination, on a mean of 3 experiments
not much differing from each other. Supposing all these facts accurate, they may be
reconciled by a few suppositions by no means
improbable. Let the atom of cerium be 22,
the protoxide 29, and the red oxide $32\frac{1}{2}$ (that
is, 1 oxy. $+$ 2 protox $= 65$); and let the
protocarbonate be 1 atom of acid, 1 of oxide,
and 1 of water; the percarbonate, 1 acid 1
oxide; the oxalate, 1 acid (40) and 1 oxide;
and the muriate, saturated with base, 3 oxide and 2 acid. Then it will be found that,

The decomposed protocarbonate will yield 57.5 red oxide;
The decomposed percarbonate will yield, 36.7 acid, 63.3
    oxide;
The decomposed oxalate will yield 47 red oxide; and
The muriate will yield 100 acid (22), and 197.7 oxide.

All of which agree very nearly with the results above obtained.

Hence it appears to me very probable that the several atoms of the metal and the oxides are as stated above; and that,

100 cerium + 31.8 oxygen = 131.8 protoxide, white.
————————+ 47.7 ——·—— = 147.7 intermediate, red.

Hisinger, from some of the same data united to other hypothetical facts than those assumed above, deduces the two oxides very different; viz. 100 metal + 17.4 oxygen for the protoxide, and 100 + 26.1 for the peroxide.

SECTION 14.

EARTHY, ALKALINE AND METALLIC
SULPHURETS.

The sulphurets exhibit a very important class of combinations of two elements. Many of the metals are found chiefly in the state of native sulphurets, and are extracted by particular processes. Artificial combinations of sulphur and the metals, and of sulphur and the earths and alkalies are commonly practised, and are found useful in chemical inves-

tigations. The alkaline and earthy sulphurets will scarcely be allowed perhaps to be
combinations of *two elements* only ; but their
analogy with the other compounds is such as
to induce us to treat of them under this head,
especially as they are agents occasionally in
the formation of metallic sulphurets, and these
cannot be so well understood without some
knowledge of the other. For like reasons
the compounds of three elements, sulphur,
metal, and oxygen, called sulphuretted oxides, and sulphuretted sulphites, and those of
four elements, sulphur, metal, oxygen and hydrogen, called hydrosulphurets, may be considered at the same time, having an intimate
relation with the sulphurets strictly so called,
or the compounds formed with sulphur and
the undecompounded bodies.

Sulphur may be combined with the earths,
alkalies and metals, by heat, of various degrees according to the nature of the subjects.
The union is attended in many cases with a
glowing ignition, indicating the evolution of
heat. The metallic oxides and sulphur when
heated together commonly produce a sulphuret of the metal, whilst the oxygen escapes
with part of the redundant sulphur in the form
of sulphurous acid, and the rest of the sulphur sublimes.

VOL. II.

In the humid way sulphur may be combined with earths, alkalies, and metals, by means of sulphuretted hydrogen, hydrosulphurets (that is, sulphuretted hydrogen united to other alkaline or earthy bases), and hydroguretted sulphurets (a name given to certain earthy and alkaline sulphurets formed mostly by boiling mixtures of the respective bases and sulphur in water.) The sulphuretted hydrogen may be used in this state of gas or combined with water; the hydrosulphurets and hydroguretted sulphurets are best applied in their watery solutions. The metals are to be used in this case in the state of salts, that is, oxides united to acids, and in solution; or their oxides may in some instances be precipitated previously to the addition of the sulphur compound; the alkalies and earths are sometimes directly sulphurized in the state of hydrates, and at other times by double affinity, in the state of salts or combined with acids. The phenomena in the case of sulphurets formed in the humid way, are various and often complicated, and the true results are not always to be obtained without considerable difficulty and uncertainty.

1. *Sulphurets of lime.*

When pounded lime and sulphur are mixed together, and heated in a crucible scarcely any union takes place; the sulphur sublimes or burns away and leaves the lime unaltered. If for lime we substitute carbonate of lime, it also remains unaltered. But if hydrate of lime and sulphur are heated together in equal weights, the hydrate is decomposed, and the lime unites to a portion of the sulphur, whilst the excess of sulphur sublimes or burns and escapes at a low red heat. The residue, about 60 per cent. of the original weight, is a yellowish white powder, composed of sulphur and lime. If this be again treated with sulphur and heated, it undergoes no material change; the last sulphur entirely escaping, leaves the sulphuret unaltered, and hence shews that it must be a true chemical compound.

Now if 32 parts hydrate of lime, which consist of 24 lime and 8 water, be mixed with 32 sulphur and heated as above, they will yield 38 parts sulphuret, which must be composed of 24 lime and 14 sulphur, or sulphur and water; but it appears from the analysis hereafter to be given, that the whole of this last part is sulphur; therefore the compound is formed of 1 atom of lime, and 1 of sulphur, and is the *protosulphuret* of lime.

When 32 parts of common hydrate of lime and 56 sulphur, are boiled together in 1000 parts water for half an hour, or more, occasionally adding water to supply the waste, a fine yellow liquid is obtained, with a few grains of residuum containing both lime and sulphur nearly in the original proportion with a few grains of alumine. This liquid of course contains in solution, a combination of 1 atom of lime, or perhaps hydrate of lime, and 4 atoms of sulphur; and may therefore be called a *quadrisulphuret* of lime. If more sulphur or lime than the above proportion be used, the surplus will remain in the residuum uncombined, shewing that by this process no other than a quadrisulphuret can be formed. A similar solution may be obtained in cold water by frequent agitation; but it is much slower in producing the effect. The strength of liquid quadrisulphuret depends upon the relative quantity of the ingredients. I have boiled it down till the water was only 5 times the other materials, which appears to be its maximum strength in the common temperature; its spe-

cific gravity was 1.146; but in general I have
used it of less than 1.07 density. It may
be proper to remark here that I find the decimals
multiplied by 4 express very nearly the number of grains of lime in 1000 grains measures
of the solution, and multiplied by 9 those of
the sulphur; on this account a solution of the
sp. gravity 1.06 facilitates the calculations, as
100 measures of it contain 2.4 grains of lime,
and 5.4 or 5.6 of sulphur nearly.

It is rather surprising that no bisulphuret
nor trisulphuret of lime should be formed this
way. One would suppose that the sulphuret of lime in its progressive changes
would have passed through the forms of bisulphuret, &c. till it had obtained its maximum
of sulphur when that was in excess; but,
as has been observed, the quadrisulphuret is
the only one formed, whatever may be the
proportions of the ingredients. I imagine
the reason to be, that the sulphur has to decompose the hydrate of lime, and that no
fewer than 4 atoms of sulphur are adequate
to that effect; it is known that water adheres
so strongly to lime as to require a red heat to
separate it. When therefore we mix lime
water with quadrisulphuret of lime, it must
be considered as a mere mixture of the two,
and that the lime does not divide the sulphur

equally. Consistently with this reasoning, whenever the lime is in excess in forming qua-drisulphuret of lime, we ought to consider the liquid solution as *lime water* holding qua-drisulphuret of lime. This d:stinction will be of some importance when the solution is weak, because then the lime in the lime water will be considerable, compared with the lime combined with sulphur.

1. *Protosulphuret.* The properties of this compound are;—about 1 grain is soluble in 1000 water; this water, as well as the powder itself, tastes like the white of an egg; salts of lead are thrown down black by the solution; weak nitric and muriatic acids dissolve the lime, and leave the sulphur · 100 parts of test acid require 19 of the powder, and yield 7 of sulphur; indicating the compound to be 12 lime and 7 sulphur. The same conclusion may be obtained by means of a solution of lead; if water containing 1.9 grains of the powder be precipitated by nitrate of lead, it will require 7 grains of the salt = 2.2 acid and 4.8 oxide, or 4.5 lead, and about 5 or $5\frac{1}{2}$ grains of sulphuret of lead will be formed, and the liquid will contain 3.4 grains of neutral nitrate of lime.

2. *Quadrisulphuret.* This combination has been long known, and some of its properties

observed; but I have not found in authors
any determination of its proportions. It is of
a beautiful yellow or orange colour, and 1
grain imparts very sensible colour to 1000 of
water; it has a disagreeable bitter taste;
when evaporated down, it crystallizes or rather perhaps solidifies into a yellowish mass;
but its properties are affected by the process
from the acquisition of oxygen. This mass
when dried, burns with a blue flame and loses
40 per cent.; the remainder is a white powder,
a mixture of sulphite and protosulphuret of
lime. Liquid quadrisulphuret exposed to the
atmosphere soon becomes covered with a white
film which arises from the sulphur displaced by
oxygen gas; this film being broken subsides,
and another is formed, and so on successively
till at length the acquisition of oxygen ceases
with the deposition of sulphur, and the liquid remains quite colourless. It is intensely
bitter, and contains lime, sulphur and oxygen
in proportions to be presently determined.
This colourless liquor undergoes a gradual
change by being kept for years in a bottle
with a common cork; a deposition of some
sulphur and sulphate of lime takes place, but
whether from a further acquisition of oxygen
gas or from some internal chemical action, I
have not had an opportunity of observing.

From the above observations it is obvious
that to form *pure* quadrisulphuret of lime the
atmospheric air should be excluded, as the
agitation by ebullition would promote the oxidizement of the compound. I mixed 168
grains of sublimed sulphur with 96 hydrate
of lime, which by previous trials I had found
to consist of 70 lime including 2 or 3 grains
of alumine, and 26 water; the mixture was
put into a small florence flask, which was
then filled with water up to the neck and
loosely corked. This was immersed in a pan
of water and boiled for 2 or 3 hours, the
flask was continually turned round to agitate
the mixture and promote the solution. After
the undissolved part had subsided the clear
liquor was decanted and found to be 2800 grain
measures of the sp. gr. 1.056; the residuum
moderately dried weighed 34 grains; it was
found to contain 8 of lime and alumine, and
25 of sulphur. Hence the liquid contained
62 lime and 143 sulphur, or 2.2 lime and 5.1
sulphur per cent.; that is, after the rate of
24 lime to 56 sulphur, or 1 atom of lime to
4 of sulphur, and its weight = 80, the atom of
sulphur being supposed 14. Here then we
have a synthetic proof of the composition being a quadrisulphuret. Innumerable other
experiments, though made with less rigid ac-

curacy, had convinced me that the liquid is essentially the same whatever the proportions of the ingredients, and that the residuum only varies in such cases.

I have made many experiments occasionally since 1805, on the quantities of oxygen absorbed and sulphur deposited by quadrisulphuret of lime. They all concur in establishing the same conclusion; namely, that each atom of the compound takes 2 of oxygen and deposits 2 of sulphur, in its transformation from the yellow to the colourless state. For instance, 100 measures of the above 1.056 took 900 of oxygen gas = 1.22 grains, and let fall 2 grains of sulphur, besides a small portion which adhered to the bottle, which was estimated at a few tenths of a grain. The method is to put 100 measures into a graduated and well stoppered bottle filled with oxygen; to agitate briskly for half an hour, occasionally opening the stopper a little under water to admit its entrance into the place of the oxygen absorbed. Whenever the agitation has been continued for five minutes without any sensible increase in absorption, and the liquor, after standing to let the sulphur subside, appears colourless, the experiment is finished. This new combination then consists of 1 atom lime, 2 sulphur, and 2 oxygen = 66; it will be necessary to give it a name: I propose calling it sulphuretted sulphite of lime, as it is an atom of sulphur united to sulphite of lime; and the rather, as it will appear in the sequel that other neutral salts do combine occasionally with an atom of sulphur. This sulphuretted sulphite may be boiled down to the sp. gr. 1.1 before it precipitates: the liquid then contains about 12 per cent. of the salt, or 5 sulphur, $2\frac{1}{2}$ oxygen, and $4\frac{1}{2}$ lime. The salt precipitates from the liquid by evaporation in the form of a white powder; it burns with a feeble blue flame, and loses about 20 per cent.; the remainder is sulphite of lime. When 100 grain measures of the liquid sulphuretted sulphite (1.1) are saturated with oxymuriate of lime, they acquire 5 grains of oxygen, and then yield $12\frac{1}{2}$ grains of sulphuric acid (containing 5 sulphur and $7\frac{1}{2}$ oxygen), as may be found by the barytic tests. The point of saturation is known by the smell of oxymuriatic acid being given out permanently.

If however we oxidize the quadrisulphuret of lime by oxymuriate of lime, the results are somewhat different. As soon as an atom of the quadrisulphuret has received two atoms of oxygen it becomes colourless as before, but $\frac{3}{4}$ of the sulphur is thrown down instead of $\frac{1}{2}$;

and when more oxymuriate is added, so as to
impart 3 atoms of oxygen to one of the salt,
a complete sulphate of lime is formed. The
point of saturation is determined by adding a
small portion of muriatic acid to the liquid,
which developes the oxymuriatic acid as soon
as it becomes in excess. This method e cels
in the analysis of the alkaline and earthy
sulphurets in general.

When quadrisulphuret of lime is treated
with an alkaline carbonate, a reciprocal change
takes place; the carbonic acid takes the lime,
and the alkali the sulphur, leaving however
1 atom of sulphur with the carbonate which
precipitates. Hence a sulphuretted carbonate
of lime is obtained and a trisulphuret of the
alkali. The sulphur burns off from the carbonate below a red heat and leaves 75 per
cent. of carbonate of lime; this affords an
excellent analysis of quadrisulphuret of lime
as far as lime is the object. Thus 540 of the
above 1.056 quadrisulphuret took 100 test
carbonate of potash (1.25), and gave a precipitate of 29 grains, which burned blue and
left 22 grains = 12 lime, and 10 acid; but if
540 : 12 :: 100 : 2.2, as above determined
synthetically: moreover, 12 lime, 10 acid, and
7 sulphur, are as 24 lime, 20 acid, and 14
sulphur; the composition of an atom of sul-

phuretted carbonate of lime, which is analogous to the sulphuretted sulphite of lime, as
found above.

When quadrisulphuret of lime is treated
with as much sulphuric acid as is sufficient for
the lime, the sulphur is in part precipitated,
but it is in union with the sulphate of lime,
or at least they are not separable by mechanical means. This compound is sold in the
shops under the name of precipitated sulphur.
It is about one half sulphate of lime, and
the other half sulphur. The nitric and muriatic acids precipitate the sulphur partially
from quadrisulPhuret, but the sulphur assumes
a viscid form and exhales sulphuretted hydrogen, and the proportion of the elements of
quadrisulphuret are not easily obtained by any
of these acids.

The mutual action of quadrisulphuret of
lime, and the metallicas lts is curious and interesting; for instance, with nitrate of lead.
Let a solution of nitrate of lead, containing
97 oxide, be treated with a solution of quadrisulphuret of lime by degrees, as long as
a blac precipitate appears, marking the exact point of saturation; this will be found when
36 parts of lime have entered, and 84 of sulphur; the sulphuret of lead will fall, and
when dried will weigh 145 parts, and contain

90 lead, and 55 sulphur; that is, 1 atom of
lead, and 4 of sulphur, and is consequently
a quadrisulphuret of lead. The liquid remains clear and colourless, and contains the
nitric acid, lime, oxygen of the lead, and $\frac{1}{3}$
of the sulphur; each atom of nitric acid combines with one of lime, which retains one
of the 4 atoms of sulphur, forming a sulphuretted nitrate of lime, consisting of 45
acid, 24 lime, and 14 sulphur; the 7 parts
of oxygen unite with 7 of sulphur to form
sulphurous acid, which require 12 parts of
lime to saturate them and 7 of sulphur, forming a sulphuretted sulphite of lime: hence
we see that 28 parts of sulphur remain in the
liquor, and the rest (56) unite with the lead.
If now we add gradually more nitrate of lead,
a silvery white precipitate appears, increasing
till half the original quantity is added, and
then the liquid is saturated. This white precipitate is sulphuretted sulphite of lead; when
heated it soon grows black and loses 15 or 20
per cent., being then a protosulphuret of lead.
The liquid now contains sulphuretted nitrate
and simple nitrate of lime; nitrate of lead
has no effect, but nitrate of mercury precipitates a black sulphuret.

Quadrisulphuret of lime saturated with oxygen, as has been observed, contains sulphu-

retted sulphite of lime in solution, and deposits sulphur: the liquid treated with nitrate of lead, gives as above the white, silvery sulphuretted sulphite of lead as a precipitate, and holds nitrate of lime in solution.

Hydrosulphuret of lime. This compound may be formed by passing sulphuretted hydrogen into lime water; the water assumes a brownish colour, but the point of saturation is not easily found, as the lime water is not neutralized so as to shew by the colour test, and water of itself absorbs above twice its volume of the gas. By means of a neutral solution of nitrate of lead it may be found that 1000 lime water in volume, require about 600 sulphuretted hydrogen, because then a mutual saturation is observed by double affinity; that is, sulphuret of lead and neutral nitrate of lime are formed; but otherwise the liquid remaining is either acid or alkaline. Hydrosulphuret of lime, as well as the other hydrosulphurets, has a peculiar bitter taste. It forms a useful reagent in regard to metals, but is apt to be spoiled by keeping, owing to the acquisition of oxygen.

2. *Sulphuret of magnesia.*

I have not succeeded in endeavouring to combine sulphur and magnesia in the dry way; but a liquid sulphuret is easily formed by the action of double affinity. Let a quantity of the liquid quadrisulphu ret of lime be treated with a solution of sulphate of magnesia, so that the sulphuric acid may be sufficient for the lime; by digesting in a moderate heat, the sulphate of lime is precipitated, carrying with it one fourth of the sulphur, and a trisulphuret of magnesia remains in solution. I have not observed any remarkable feature of distinction between this sulphuret and that of lime, except as above noticed in the proportions of their compounds.

Hydrosulphuret of magnesia. This compound may be formed by pouring sulphuret ted hydrogen water into recently precipitated magnesia; it does not differ much from that of lime. One atom of sulphuretted hydrogen (15), combines with one of magnesia (17), and the compound is soluble in water.

3. Sulphuret of barytes.

Protosulphuret. The protosulphuret of barytes may be procured the same way as that of lime, by heating hydrate of barytes and sulphur till the mixture becomes red. It is very little soluble in water, and accords in other respects with the like compound of lime. It consists of 68 barytes and 14 sulphur, or 100 barytes and 20½ sulphur.

Quadrisulphuret. The quadrisulphuret of barytes may be formed the same way as quadrisulphuret of lime, by boiling the hydrate of barytes and sulphur together. A yellow solution of the compound is formed, not distinguishable in appearance from that of lime; and it appears to be analogous to it in most of its properties. By acquiring oxygen it becomes colourless sulphuretted sulphite of barytes, and crystalizes in needles; in this last respect it differs from that of lime. The maximum density of liquid quadrisulphuret I have not had an opportunity of ascertaining; it is 1.07 or upwards; that of the liquid sulphuretted sulphite is much less than that of lime; the crystals are found in a liquid so low as 1.004 sp. gr. They have a fine silky lustre when dry, and a yellowish colour; heated they burn with a blue flame and leave a white mass of sulphate preserving the same crystalline appearance as before, and lose about 20 per cent. of weight. Ten grains of the crystals of sulphuretted sulphite, when treated with liquid oxymuriate of lime to saturation, require 2+grains of oxygen and yield 8 grains of sulphate of barytes, together with an excess of sulphuric acid which with muriate of barytes gives 8 grains more of sulphate. From these facts it may be concluded that the sulphuretted sulphite consists of one atom barytes, 2 sulphur, 2 oxygen, and 2 water, and that 4 more of oxygen are derived from the oxymuriatic acid to convert the sulphurous oxide into sulphuric acid. The sulphuretted sulphite of barytes seems to pass into sulphate by length of time. The weight of the atom of quadrisulphuret of barytes is 124; the compound in mass consists of 100 barytes and 82 sulphur.

Hydrosulphuret of barytes. This compound may be formed in the same manner as that of lime, and is found to have similar properties. The proportions for mutual saturation are, I find, as in the case of lime, 15 sulphuretted hydrogen to 68 barytes by weight, or one atom of each.

4. *Sulphurets of strontites.*

The protosulphuret and quadrisulphuret of strontites may be formed in the same way as those of lime and barytes. From a few experiments made on these compounds I have not observed any remarkable feature of distinction between them and the corresponding ones of the other earths.

Hydrosulphuret of strontites. This compound may be formed in the same way as that of lime; the proportions to produce mutual saturation will be 1 atom of each, or 15 parts sulphuretted hydrogen, to 46 strontites by weight.

5, 6, 7, 8, *and* 9. *Sulphurets of alumine, silex, yttria, glucine,* and *zircone.*

I made several unsuccessful attempts to combine alumine and sulphur. When alumine and sulphur mixed together are heated, the sulphur sublimes chiefly, and leaves the alumine with traces of sulphate of alumine.

In the humid way, recently precipitated and moist alumine mixed with sulphur and

boiled in water, give a liquid with some traces of sulphuric acid, but no sulphuret of
alumine; the sulphur and alumine both subside, and when the sulphur is either sublimed
or burnt, the alumine remains much the same
as at first. When a solution of alum is
treated with sulphuret of lime, sulphate of
lime is precipitated along with the greatest
part of the sulphur in a kind of feeble union
rather than mechanical mixture, it should
seem; the alumine is at the same time precipitated probably in mechanical mixture;
there remain in solution a little sulphuret of
potash and sulphate of lime.

Sulphuret of silex is not known, I appre-
lhend, to exist. When silicated potash in so-
ution is treated with quadrisulphuret of lime,
a copious dark brown or black precipitate instantly appears; the liquid when filtered is
of a pale yellow colour, and seems to contain
about one half of the sulphur and potash,
whilst the other half is thrown down in
union with the lime and silex. This black
compound is probably 1 atom of lime, 2 of
sulphur, 2 of potash, and 2 of silex; it cannot therefore be accounted a sulphuret of
silex.

Sulphurets of yttria, glucine, and *zircone,*
are as yet, I presume, unknown.

10. *Sulphurets of potash.*

Potash has a strong affinity for sulphur and unites with it in various ways and proportions.

1st. *In the dry way by heat.* When either pure potash or the carbonate (salt of tartar) is heated in a covered crucible with sulphur, a chemical union of the two principles takes place. Eight parts of dried hydrate of potash unite to six or seven of sulphur: a heat of 4 or 500° of Fahrenheit is convenient for the purpose. If the carbonate of potash be used, then 12 parts dried in a low red heat will require 8 of sulphur for their complete saturation: in this case a higher degree of heat is requisite in order to expel the carbonic acid; a low red heat seems sufficient from my trials. When the heat does not exceed 3 or 400° a partial union takes place; the carbonate of potash, without losing any acid, unites to $\frac{1}{3}$ of the sulphur, and the rest of the sulphur remains uncombined; when intermediate degrees of heat are used, I have found the result a mixture of the pure sulphuret and the carbonated sulphuret, with more or less of sulphate of potash. A high degree of heat and exposure to the atmosphere produces a sulphate instead of a sulphuret. The sulphurets obtained this way are in fusion till poured out and cooled; they are of a *liver colour*, and hence were formerly called *livers of sulphur*. They are largely soluble in water, and give a brownish yellow solution.

2d. *In the humid way by solution.* Pure caustic potash in solution when boiled with sulphur dissolves it largely, 42 parts of real potash being satuiated with about 56 of sulphur. If we boil a solution of carbonate of potash with sulphur, for an hour or more, a brown liquor is obtained, which consists of 60 parts carbonate of potash and 14 sulphur in chemical union.—It has already been observed that a trisulphuret of potash may be obtained by double affinity from quadrisulphuret of lime and carbonate of potash, together with sulphuretted carbonate of lime.

From what has been stated we may infer at least three varieties in the compounds of sulphur and potash, viz.

1st. *Sulphuretted carbonate of potash.* This consists of 1 atom carbonate of potash (61) with I atom of sulphur (14). Its analysis may be effected as follows : the quantity of carbonic acid may be found by the lime water necessary to saturate it ; the potash may be known from the quantity previously entering into the mixture; and the sulphur in the same manner, or from the quantity of sulphuretted carbonate of lead that it forms.— The sulphur may also be known, from the quantity of oxygen it requires by means of oxymuriate of lime to produce saturation; this I find to take place when the oxygen is half the weight of the sulphur, or one atom to one of sulphur; it soon happens, that one atom of sulphur deprives two others of their oxygen, and sulphuric acid is formed whilst the other two atoms of sulphur join the carbonate of lime and are precipitated along with it. As it may frequently happen, that the sulphuretted carbonate is mixed with common carbonate of potash, the proportions may be found by means of nitrate of lead, which being cautiously dropped into the solution, lets fall first the brown sulphuretted carbonate of lead, and then the common white carbonate of lead.

The sulphuretted carbonate of potash absorbs oxygen and precipitates metals much the same in appearance as the other sulphurets; but essential distinctions are observable, some of which are noticed above, and others will appear in the sequel.

2 and 3. The *trisulphuret* and *quadrisulphuret* of potash so nearly resemble the qua-

drisulphuret of lime in their properties, as
not to require any additional remarks.

Hydrosulphuret of potash. This combination, when duly proportioned, consists of 15
parts sulphuretted hydrogen, and 42 potash
by weight, or one atom of each. It may be
formed by directly uniting the two elements,
or by decomposing hydrosulphuret of lime by
carbonate of potash. Its properties agree
with those of the other hydrosulphurets.

11. *Sulphurets of soda.*

I have repeated most of the experiments
on the sulphurization of potash with soda, and
have not found any one remarkable feature of
distinction, besides those which arise from
the weights of the atoms.

1. *Sulphuretted carbonate of soda* consists
of 1 atom of carbonate of soda united to 1
of sulphur; or of 47 parts of the former and
14 of the latter.

2. *Trisulphuret of soda* consists of 1 atom
soda (28) and 3 of sulphur (42).

3. *Quadrisulphuret of soda* consists of 1
atom soda (28) and 4 atoms of sulphur (56).

Hydrosulphuret of soda. This compound
consists of one atom of each of the elements

or 15 sulphuretted hydrogen, and 28 soda. In other respects it agrees with hydrosulphuret of potash.

12. *Sulphuret of ammonia.*

The best way which I have found of procuring sulphuret of ammonia, is to treat quadrisulphuret of lime with the carbonate of ammonia as long as any precipitate takes place; the precipitate is sulphuretted carbonate of lime, 3 atoms of sulphur to 1 of carbonate of lime. The liquid is of a pale yellow, and contains ammonia and sulphur united in the ratio of 1 atom (of 6) to 1 of sulphur: it may therefore be denominated the protosulphuret of ammonia.

The carbonate of ammonia is best procured by heating the common subcarbonate of ammonia, first pulverized, in a temperature of 100° for half an hour, or exposing it for a few days to the atmosphere. What remains of the salt is almost without smell; it should consist of 19 parts acid, 6 ammonia, and 8 water nearly: the ammonia is usually however a small degree in excess.

Hydrosulphuret of ammonia. This compound may be formed in the dry state by combining the two gases of sulphuretted hydrogen and ammonia over mercury; it is of a white crystalline appearance, and very soluble in water, and forms a fuming liquor of a very pungent smell. It may also be obtained by passing sulphuretted hydrogen into a vessel containing liquid ammonia. I find about 110 or 120 measures of sulphuretted hydrogen require 100 of ammoniacal gas. Hence it is 1 atom of sulphuretted hydrogen (15), that unites to 1 of ammonia (6).

13. *Sulphurets of gold.*

There exist at least two sulphurets of gold, the nature and proportions of which are easily ascertained; though several authors assert that no combinations of gold and sulphur are known; amongst these it is surprizing to find Proust : indeed most of the others have probably been led by his authority to adopt the opinion without examination. It is not very easy to account for his deception.

Obercampf, in the Annal. de Chimie, tom. 80. 1811, is the first author I have seen who distinctly maintains the existence of one or more sulphurets of gold, though it seems to have been admitted previously by Bucholz. The last author finds 82 gold unite to 18 sulphur, and the former 80 to 20 nearly.

Protosulphuret of gold. This compound is formed whenever a solution of muriate of gold is agitated with sulphuretted hydrogen gas, or with the same united to a base, as lime or alkali. A black or deep brown powder falls down by the addition of more gas, till the whole of the gold is precipitated. The oxide of gold loses one atom of oxygen, and receives one of sulphur in its place, whilst the hydrogen of the gas is carried off along with the oxygen of the oxide. The sulphuret dried and heated, burns with a blue flame, leaving the gold nearly pure. This compound consists, I find, of 81 gold and 19 sulphur per cent.; or 100 gold unite to 23 sulphur.

Trisulphuret of gold. This compound is obtained whenever quadrisulphuret of lime is gradually dropped into a solution of muriate of gold; it is a black powder, not quite so deep as the former. Care must be taken to saturate the excess of acid previously by lime-water, to prevent any uncombined sulphur precipitating. Trisulphuret of gold being heated, burns with a blue flame, and leaves the gold nearly pure; it loses from 10 to 45 per cent. by the process. It is constituted of 1 atom gold and 3 sulphur, or 60 gold

and 42 sulphur, nearly; or 100 gold combine
with 70 sulphur

From several experiments I am led to conclude that each atom of oxide of gold takes
3 of sulphur, and parts with 1 of oxygen to
the remaining sulphur; thus a trisulphuret of
gold is formed, and an oxide of sulphur; the
liquid, being afterwards treated with oxymuriate of lime, is found to require twice the
oxygen of the gold for its saturation, when
a corresponding portion of sulphuric acid may
be precipitated by muriate of barytes.

14. *Sulphuret of platina.*

Sulphur may be combined with platina in
several ways, and probably in different proportions; but the combination is not so easily
and elegantly effected as with many other metals, and hence some uncertainty still remains
on the subject.

When a salt of platina is treated with sulphuret or hydrosulphuret of lime, or sulphuretted hydrogen water, the liquid slowly and
gradually grows dark brown and finally black;
after agitation and standing a few hours, the
liquid is semitransparent, and a black flocculent precipitate appears at the bottom. Sometimes after violent agitation. the liquid on standing a few minutes becomes a transparent brown.

but soon grows turbid again. In the course of a few days, and by occasional agitation, the liquid finally becomes clear and nearly free from platina, and the precipitate may be collected on a filter and dried. This circumstance of slow and indolent precipitation cannot be prevented by any means I have found, such as saturating the excess of acid, &c.

Mr. Edmund Davy, in the 40th VOL. of the Philos. Magazine, has given us the results of his experiments and observations on the sulphurets of platina, containing some useful and original information. He combines platina with sulphur by heating the ammonia-muriate of platina with sulphur; also by heating platina and sulphur in an exhausted tube; and by sending sulphuretted hydrogen gas or water into a solution of muriate of platina; this precipitate he calls hydrosulphuret of platina.

He has just noticed the precipitate formed by sulphuret of potash with muriate of platina, but gives no opinion as to the compound obtained this way. He determines three sulphurets, namely,

Subsulphuret,   100 platina + 19 sulphur
Sulphuret,   100   ——   + 28.2 ——
Supersulphuret,   100   ——   + 38.8 ——

I have obtained the sulphuret of platina in five ways: 1st. By pouring sulphuret of lime solution by degrees into muriate of platina, and agitating the mixture well or till it grew black each time; after digesting for some days, repeated filtering, and drying, a black powder is obtained: 2. Instead of sulphuret, hydrosulphuret of lime was used; the precipitate was obtained under like circumstances: 3d. Sulphuretted hydrogen water was used, and the precipitate obtained in like manner : 4th. Ten grains of ammonia-muriate of platina were treated with sulphuretted hydrogen water; by continued agitation the yellow powder disappeared, the liquid looked uniformly black, and at length a precipitate was formed; by repeated filtration and addition of sulphuretted hydrogen water, the whole of the platina was thrown down, and the liquid remained colourless; but it is difficult to discover the exact quantity of sulphuretted hydrogen requisite for any weight of the ammonia-muriate from the tediousness of the operation; 6 grains of well dried black powder were obtained, besides perhaps 1 grain of loss on the filters: 5th. Ammonia-muriate of platina was heated in a covered crucible along with sulphur till it was judged that all the uncombined sulphur was sublimed or dissipated.

All these sulphurets appear to me to be the

same when dried in a moderate heat. When exposed to a low red heat they yield water and sulphurous acid, and lose about $\frac{2}{5}$ of their weight.

The subject however, requires further investigation. The sulphurets of platina appear of a complex nature, and the proportions of their elements are not yet determined with precision.

15. Sulphurets of silver.

Silver combines with sulphur in two different proportions, and forms two sulphurets, both of them black or dark brown.

1. *Protosulphuret of silver.* This may be formed either by the dry or humid way: if thin lamina of silver be heated with sulphur, they combine and form this sulphuret; a higher degree of heat expels the sulphur again. It is formed too by passing sulphuretted hydrogen or a hydrosulphuret through a solution of silver in nitric or other acids. The atom of silver unites with that of sulphur, whilst the hydrogen unites with the oxygen. Of course this compound is composed of 90 silver, and 14 sulphur, and the atom weighs 104; or 100 silver unite with 15.5 sulphur. Klaproth finds 100 silver and 17.6 sulphur; Wenzel 100 silver, and 14.7 sulphur; Ber-

zelius 100 silver, and 14.9 sulphur; and Vauquelin 100 silver, and 14 sulphur.

Trisulphuret of silver. This compound
is formed whenever neutral nitrate of silver
is dropped into a solution of quadrisulphuret
of lime or alkali. Mutual saturation seems
to take place when eight atoms of nitrate
meet with seven of quadrisulphuret. Trisulphuret of silver is constituted of 90 silver,
and 42 sulphur; or of 100 silver, and 46.5 sulphur. Its colour is not so dark as that of the
protosulphuret. The residuary liquid contains sulphurous acid, which is easily converted into sulphuric by the addition of a portion of lime; and the quantity of acid may
then be determined by muriate of barytes.

16. *Sulphurets of mercury.*

Mercury combines readily with sulphur
both in the dry and humid way, and that in
several proportions, as under: namely,

1. *Protosulphuret of mercury.* This is most
conveniently formed by passing sulphuretted
hydrogen gas through a solution of the protonitrate of mercury, or by pouring hydrosulphuret of lime, &c. into the same solution.
The protosulphuret falls down in the state of
a black powder It consists of 167 mercury,

and 14 sulphur; or of 100 mercury, and 8.4
sulphur. The theory of its formation is the
same as that of silver.

2. *Deutosulphuret of mercury* This is formed in the humid way whenever sulphuretted
hydrogen or a hydrosulphuret in excess is mixed with the deutonitrate or deutomuriate of
mercury (corrosive sublimate); a brown powder is precipitated which is the deutosulphuret. If the sulphuretted hydrogen be only one
half what is sufficient to form the deutosulphuret, then we obtain no sulphuret, but instead of it a protonitrate or protomuriate, as
was first intimated by Proust; I find however,
the atom of sulphur adheres to the atom of
salt, and that it is therefore a sulphuretted
protonitrate or muriate, whilst 1 atom of
oxygen unites with the hydrogen. The brown
precipitate does not change to yellow, orange,
and red, when left undisturbed for a few days,
in my experience; though this is stated to
have been observed by Mr. Accum. Notwithstanding the difference in colour, this
deutosulphuret must be the same nearly as the
cinnabar and vermillion of commerce, if
Proust and others are right in their analysis
of these articles. The combination of the elements of sulphur and mercury when intended
to form cinnabar is made in the dry way by tri-

turation, and a moderate heat; the compound, at first black, is afterwards sublimed by a duly regulated heat and becomes red. This compound must consist of 100 mercury and 17 sulphur nearly.

3. *Quadrisulphuret of mercury.* This compound is formed when a solution of protonitrate of mercury is treated with quadrisulphuret of lime, added by degrees till the clear liquid no longer gives a dark coloured precipitate. The oxygen of the mercurial salt unites, it should seem, to part of the sulphur, and forms sulphuric acid, whilst the rest of the sulphur unites to the mercury. This sulphuret is a black or dark brown powder, and when heated burns with a blue flame. It consists of 100 mercury, and 33 or 34 sulphur, as appears to me from the synthesis.

When the insoluble muriate of mercury (calomel), is triturated in liquid quadrisulphuret of lime, it is soon decomposed; quadrisulphuret of mercury is formed, with muriate of lime and sulphuric or sulphurous acid.

When the soluble muriate (corrosive sublimate), has quadrisulphuret of lime dropped into it by degrees; at first a yellowish white precipitate is obtained, which increases till it is one half saturated; after this, by continually adding more sulphuret, the precipitate grows darker, and ends in being quite black. It is at least as high as quadrisulphuret. Much sulphurous acid is found in the liquid.

The deutonitrate of mercury, produces a copious yellow precipitate with quadrisulphuret of lime. Exposed to the sun, it grows black in a few minutes on the light side, but continues yellow on the opposite side of the jar; at the same time, an effervescence and disengagement of oxygen gas are observed. Finally, the precipitate becomes the common quadrisulphuret, and the liquid contains sulphurous and sulphuric acids.

The recently precipitated and washed oxides of mercury act upon quadrisulphuret of lime. The black oxide seems to take 4 atoms of sulphur and part with its oxygen to another portion of sulphur; but the red oxide becomes light brown and retains the colour when dried. It seems to take the same sulphur as the black, but whether it retains any of the oxygen, I have not ascertained. The action is more slow than when the nitrates are used, and more quadrisulphuret of lime is expedient.

Mercury and sulphur combine in the dry way by trituration and by heat, forming a black powder; but the species of compounds

and quantities of the ingredients combining
in this mode, have not been ascertained.

17. *Sulphuret of palladium.*

Berzelius exposed 15 grains of palladium
filings mixed with as much sulphur to a heat
sufficient to expel the uncombined sulphur.
The increase of weight was 28 per cent. upon
the palladium; when exposed afresh with sulphur to heat, no addition was made to the
weight.

Vauquelin heated 100 parts of the triple
salt of palladium with an equal weight of sulphur, and obtained 52 parts of a blueish white
sulphuret, very hard, and when broken ex-
hibiting brilliant plates in its fracture. He
had previously found that 100 salt contained
40 to 42 of metal: hence 100 metal combined with from 24 to 30 of sulphur. This
agrees nearly with the above results of Berzelius. A very high degree of heat expels the
sulphur and oxidizes the metal; but a more
moderate heat leaves the palladium of a silver
white colour and nearly pure. According to
this the atom of protosulphuret of palladium
must consist of 50 palladium, and 14 sulphur.

18. *Sulphuret of rhodium.*

Vauquelin found that 4 parts of the ammonia-muriate of rhodium (containing 28 or 29
per cent. of metal) being mixed with an equal
weight of sulphur, and heated, a blueish
white button was obtained, weighing 1.4.
Hence 100 metal seem to take 25 of sulphur;
and allowing this to be the protosulphuret of
rhodium, the atom must consist of one rhodium
56, and one sulphur 14, making the whole
weight 70.

19. *Sulphuret of iridium.*

According to Vauquelin, 100 parts of the
ammonia-muriate of iridium heated with as
much sulphur, yield 60 parts of black powder
resembling the other metallic sulphurets; but
100 parts of the salt were found to yield from
42 to 45 of metal. Now supposing the last
number the most correct, it should seem that
3 parts iridium take 1 sulphur, or 100 take 33⅓
This being supposed the protosulphuret, the
atom of iridium must be 42, and that of the
sulphuret 56.

20. *Sulphuret of osmium.*

It is as yet unknown whether any combination of sulphur and osmium exists.

21. *Sulphurets of copper.*

Copper readily unites with sulphur both in the dry and humid way When 3 parts of copper filings are mixed with 1 part of sulphur, and heat applied, a brilliant combustion ensues, which indicates the union of the two bodies. Copper leaf burns in the fumes of sulphur, as Berzelius has observed, with great brilliancy.

The protosulphuret of copper obtained by these similar methods, when pulverized, is black or dark coloured; it has been analyzed by various authors, who nearly agree in their results. Proust finds 100 copper unite with 28 sulphur; Wenzel, 100 copper and 25 sulphur; Vauquelin, 100 copper and 27 sulphur; and Berzelius 100 copper and 25 sulphur.

If the atom of copper be 56, and that of sulphur 14, the atom of protosulphuret of copper will be 70; which gives just 100 copper and 25 sulphur.

The protosulphuret may also be formed in the humid way, by sending sulphuretted hydrogen gas or a hydrosulphuret into a solution of protomuriate of copper, or amongst the recently precipitated protoxide of copper.

Deutosulphuret of copper. This compound is formed whenever sulphuretted hydrogen gas

or a hydrosulphuret is sent into a solution of
salt containing the deutoxide, or into the
deutoxide just precipitated from any acid. It
is a dark brown powder not differing much in
appearance from the protosulphuret. It consists of 100 copper and 50 sulphur; the
weight of the atom is 84.

Quadrisulphuret of copper. This compound
is formed by mixing quadrisulphuret of lime
with a salt of the deutoxide of copper, and
diluting the solution.   A light brown precipitate falls immediately, which is the quadrisulphuret of copper. It burns with a blue
flame, and leaves the protosulphuret. The
atom consists of 56 copper and 56 sulphur,
or weighs 112; and hence the sulphuret consists of equal parts copper and sulphur.

The blue hydrate of copper recently precipitated from a salt of copper and washed, acts
upon quadrisulphuret of lime; the results, according to my experience, is quadrisulphuret
of copper, and the oxygen unites with the
sulphur remaining in the liquor.

22. *Sulphurets of iron.*

Sulphur may be united to iron either by the
dry or humid way, and that in various proportions.

Protosulphuret of iron. This compound may
be formed by passing a hydrosulphuret into a
solution of the green oxide in any acid. It
is a black powder. It may also be formed by
rubbing a highly heated bar of iron with a
roll of sulphur; the two unite in a fluid form
and soon congeal into a brownish black mass.
It is too a natural production, though not
very common; excellent analyses of it, as
well as of the common pyrites, were some time
ago given by Mr. Hatchett. (See Nicholson's
Journ. Vol. 10.) The protosulphuret is magnetic in a considerable degree; it is soluble in
acids, and yields sulphuretted hydrogen. It
is proper to notice that the sulphuret of iron
is not precipitated from solutions by sulphuretted hydrogen simply or without a base.
According to Mr. Hatchett this sulphuret
consists of 100 iron, and 57 sulphur, which
corresponds with 1 atom iron 25, and 1 of sulphur, 14, nearly.

Deutosulphuret of iron. This is a natural
production frequently met with, and in various forms; it is called pyrites, or iron pyrites;
it is a yellowish mineral and often appears
when broken, of a radiated texture, but
sometimes it is crystallized in cubes or do-
decaedrons. Acids have little effect upon it,
except the nitric, which when diluted attacks

both the sulphur and iron; much nitrous gas
is produced, the iron is dissolved, and the
sulphur chiefly converted into sulphuric acid.
This sulphuret consists, according to Proust,
of 100 iron, and 90 sulphur, and with this
Bucholz recently agrees (Nichols. 27—356);
but Hatchett makes it 100 iron, and 112 sulphur. From an experiment of my own on
the radiated pyrites, I found nearly equal parts
of iron and sulphur. One atom of iron (25,)
and two of sulphur (28,) would give 100 to 112;
but if the atom of sulphur be only 13, it gives
100 iron to 104 sulphur. Mr. Hatchett un-
fortunately calculating the proportions of the
ingredients *in* 100 sulphuret, instead of *on*
100 iron, did not notice that the sulphur in the
common pyrites is just double of that in the
magnetic pyrites.

Quinsulphuret of iron. This combination
consisting of 5 atoms of sulphur with 1 of iron,
is formed by mixing a solution of green sulphate of iron with quadrisulphuret of lime in
due proportion. I found 50 measures sulphate
1.168 saturate 310 of 1.05 sulphuret diluted
so as to become 6 oz.; this yielded 14 grs. dried
sulphuret of iron = 3.6 iron, known to be in
the sulphate, and 10.4 sulphur the liquid contained 2+ sulphur combined with the lime and
oxygen of the oxide; for it took 2.3 oxygen

by means of oxymuriate of lime to convert
the sulphur into sulphuric acid together with
1 + from the oxide, making 3 + oxygen,
which unites to 2+ sulphur to constitute 5+
sulphuric acid; and this quantity of acid was
found to exist by muriate of barytes together
with five more brought in by the sulphate of
iron. This sulphuret is a yellowish brown
powder; it readily exhales sulphur by heat and
is reduced to the protosulphuret; but in the
open air it burns with a blue flame and leaves
the protosulphuret partially, as I apprehend,
oxidized. The theory of the formation of
quinsulphuret seems to be this: 3 atoms of
quadrisulphuret of lime are requisite to saturate 2 of sulphate of iron; the 2 atoms of sulphuric acid seize 2 of lime, three fourths of
the sulphur unite to the iron, and one fourth
to its oxygen, forming 2 atoms of oxide of
sulphur, which attack the 3d atom of sulphuret and decompose it, giving its sulphur
to the iron, and neutralizing the lime (for
the liquid is found neutral) In this way 10
atoms of sulphur are united to 2 of iron,
and 2 of sulphur to 2 of oxygen, with one of
lime, which last compound remains in solution,
and the oxide of sulphur may be converted into sulphuric acid immediately by the
application of oxymuriate of lime.

It is remarkable that neither the green nor the yellow oxides of iron, even when recently precipitated and not dried, seems capable of decomposing quadrisulphuret of lime.

It is probable that trisulphuret and quadrisulphuret of iron may be formed; but I have not ascertained the truth of this opinion.

23. Sulphurets of nickel.

Protosulphuret. According to Proust, nickel unites to sulphur by heat, so that 100 take 46 or 48; the sulphuret is of the colour of common pyrites. (Journ. de Physique, 63 and 80). According to Mr. Ed. Davy 100 nickel take 54 sulphur. By saturating a solution of nitrate of nickel with hydrosulphuret of lime I obtained 40 grains from 33 protoxide or 26 metal. This was evidently the protosulphuret; it was a fine black powder, and consists of 100 metal and 54 sulphur.

Quinsulphuret. This compound may be obtained from nitrate of nickel and quadrisulphuret of lime, in the same manner as that of iron. It is a deep black powder, and consists of 100 nickel, and 215 sulphur. By exposure to heat, the greatest part of the sulphur burns off, and the rest may be expelled by an increase of temperature.

Probably intermediate sulphurets may be

formed; but I have not pursued the investigation.

24. *Sulphurets of tin.*

Sulphur and tin unite both in the dry and humid way, and in various proportions.

Protosulphuret. This may be readily formed in the dry way as follows; let 100 grains of tin be fused in a small iron ladle and heated to 6 or 8 hundred degrees Fahrenheit; let then small pieces of sulphur of 10 or 20 grains be successively dropped into the fused metal: a copious blue flame will instantly arise each time, and a glowing heat will take place, when the sulphur and tin are in contact; as soon as this ceases, another fragment of sulphur must be dropped in, and this two or three times repeated, heating it at last to a perfect red; the mass may then be taken out and pounded in a mortar; a great part of it will be a pulverulent powder, but some portions of malleable metal will still be mixed with it, which may be separated by a sieve. This must be again heated and treated with sulphur as before, and the whole mass will be converted to a sulphuret. I find that 100 parts of tin become in this way 127 grains; which is the due proportion of 52 tin and 14 sulphur, so that no loss of tin is sustained by the process when duly managed. According to Wenzel, 100 tin take 18 sulphur; Bergman, 25; Pelletier, 15 to 20; Proust, 20; but Dr. John Davy and Berzelius find nearly 27 as above stated, and I have no doubt it is near the truth.

The protosulphuret of tin is a dark grey shining powder, with a streak like molybdena or plumbago; it is not much different in colour and appearance from native sulphuret of antimony, only less blue. It is soluble in muriatic acid by heat, and yields sulphuretted hydrogen and protomuriate of tin.

Deutosulphuret. This compound is better known than the former: it may be formed in various ways; one is by heating a mixture of deutoxide of tin and sulphur in a retort almost to a red heat; sulphur sublimes and sulphurous acid is disengaged, and there remains a yellow, light shining, flaky mass at the bottom of the retort which is the sulphuret. It was formerly called *aurum musivum* or mosaic gold. Pelletier and Proust were of opinion that this product is a sulphuretted oxide of tin; but Dr. John Davy and Berzelius have rendered it more probable that it is a true deutosulphuret, consisting of 100 tin and 54 sulphur. It is insoluble in muriatic or nitric acid, but slowly soluble by the compound of the two acids; it is also soluble in potash by heat. By exposing it to a bright red heat, it burns with a blue flame and leaves a yellowish powder which does not seem to differ much from protosulphuret.

Berzelius distilled a mixture of protosulphuret and sulphur at a low red heat, and obtained a mass of a yellow grey colour and metallic lustre, which consisted of 100 tin, and 14 sulphur, which is just the mean sulphur between the other two. This would seem to indicate that a compound of the two sulphurets, 1 atom to 1, is capable of being formed.

Hydrosulphuret of tin minor. This compound is formed according to Proust, when sulphuretted hydrogen, or an alkaline or earthy hydrosulphuret is passed into a solution of protomuriate of tin. It is of a brown or dark coffee colour when precipitated, and black when dried. By heat it yields water and protosulphuret. From some experiments I am inclined to believe, that it is formed of 1 atom protosulphuret and 1 of water: or, which is the same, 1 atom protoxide of tin and 1 of sulphuretted hydrogen. If this be right it may be said to be a compound of 100 tin, 27 sulphur and 15 water.

Hydrosulphuret of tin major. This name is

given by Proust to the yellow compound
thrown down by sulphuretted hydrogen or by
hydrosulphurets from solutions of the deutoxide of tin. When dried moderately, the
precipitate is of a dull yellow colour, and vitreous fracture, but I find it is almost black,
dried in a heat of 150° or upwards. By moderate heat it yields water, sulphurous acid,
sulphur, and the residue is deutosulphuret of
tin according to Proust. I heated 4 parts of
the above previously dried so as to become a
black vitreous powder; it burned feebly with
a blue flame, and after being made moderately red, left nearly 3 parts exactly resembling the artificial protosulphuret. I believe
the dried precipitate will be found to be constituted of 1 atom tin, 2 sulphur and 1 water;
that is, 100 tin, 54 sulphur and 15 water
= 169 by weight; and that it loses 27 sulphur
and 15 water by a red heat, which reduces the
weight just one-fourth.

Quinsulphuret of tin. This is obtained in
the humid way, by first precipitating the oxide, and then putting quadrisulphuret of lime
or potash to the liquid containing the precipitate, till the liquid after agitation and subsidence of the precipitate continues of a yellowish colour. I found that 31 measures of
protomuriate of tin of 1.377 = 7 grains acid,

7.5 tin and 1 oxygen, precipitated by 10 oz.
lime water, required 450 measures of 1.40
sulphuret of lime, containing 16 sulphur and
7.2 lime, for their saturation. The residuary
liquid was nearly colourless, and the precipitate dried in an oven of 100° or more, for 10
hours, weighed 17 grains besides loss in the operation. It was a yellow, vitreous mass, and
when pulverized and heated, burned with a
blue flame, and lost 40 per cent. in weight;
the residue was a yellow grey colour, and
seemed to be like the intermediate sulphuret
of Berzelius; it would not give sulphuretted
hydrogen by hot muriatic acid. Now if 52
(1 atom tin) : 70 (5 atoms sulphur) : : 7.5
tin : 10 + sulphur; hence the sulphuret should
have weighed 17.5 grains, which was the
observed weight, allowing ½ grain for loss.
According to this, 100 tin combine with 135
sulphur, and when burnt, the 235 are reduced to 140, the weight observed by Berzelius
in the instance alluded to. The liquid required 5 grains of oxygen from oxymuriate of
lime, to convert the sulphur into sulphuric
acid, and the weight of this acid, found by
muriate of barytes, was 11 grains, indicating
4.4 sulphur. It may be observed that the 4.4
grains, and 10 grains, do not make up the
whole (16) of the sulphuret of lime; but the

reason I apprehend was, that the quadrisulphuret was old, and did not contain the full
share of sulphur, it being usual for a small
part to fall by time.

The deutomuriate of tin, precipitating the
oxide in like manner, yielded a sulphuret rather lighter yellow than the above; about 10
tin gave 25 grains of sulphuret dried in a
temperature of 80 to 100°. This compound
still contained water, and I suspect it will be
found constituted of 1 atom tin, 5 sulphur,
and 2 water.

25. *Sulphurets of lead.*

Lead combines with sulphur in various proportions, some of which are natural produc-
tions of great purity.

Protosulphuret. This is a natural production which is called galena; it is of lead grey
colour and metallic appearance, and is found
both in masses and crystallized; its sp. gr. is
about 7.5. It may be formed artificially by
heating lead or its oxide with sulphur; also by
treating a solution of lead with sulphuretted
hydrogen or with a hydrosulphuret. Authors
are well agreed as to the proportions of the
ingredients; 100 lead combine with from 15
to 16 sulphur. That is, 90 lead with 14 sulphur; or 1 atom of lead with 1 of sulphur.

Deutosulphuret. Dr. Thomson mentions a natural production or species of galena which contains twice the quantity of sulphur of that above. I have reason to believe that this compound is easily formed in the humid way, by treating the precipitated oxide with the due quantity of quadrisulphuret of lime.

Trisulphuret and *quadrisulphuret.* These compounds, I find, may be formed by means of quadrisulphuret of lime or potash. When a solution of any salt of lead or the recently precipitated and moist oxide, is treated with the requisite quantity of quadrisulphuret of lime, a combination consisting of 1 atom of lead and 3 of sulphur is formed. It is a black powder not differing much in appearance from the protosulphuret; it is lighter and more spongy. It consists of 100 lead and 46 or 47 sulphur. The due proportions of the elements to form the above compound are, lead 100 parts in solution, and sulphur, 62 parts; ¼ of the sulphur is retained by the lime, and may be converted into sulphuric acid instantly by the addition of as much oxymuriate of lime as contains oxygen equal in weight to the sulphur, as it has already as much oxygen as converts it into sulphurous oxide, derived from the oxide of lead.

Quadrisulphuret of lead is to be obtained in

the same way; only we must have an excess
of the sulphuret of lime, or more than 80 sulphur for 100 lead in solution, as $\frac{1}{5}$ part of the
sulphur at least is retained by the lime. The
quadrisulphuret is a black powder like the
others; it burns with a blue flame and loses
nearly 40 per cent., the residue being still
black. It consists of 100 lead and 62 sulphur.

I have not ascertained whether any higher
sulphuret of lead is capable of being formed
this way.

It has been already noticed (page 109),
that a beautiful white, silvery sulphuretted
sulphite of lead is formed and gradually precipitated, when nitrate of lead is dropped into
a solution where as much black quadrisulphuret of lead has been just thrown down as the
sulphuret of lime can form.

26. Sulphurets of zinc

Zinc and sulphur are scarcely to be united
directly by heat; but by heating the oxide of
zinc and sulphur together, a combination is
effected; part of the sulphur carries off the
oxygen in sulphurous acid, and part combines
with the zinc. Mineralogists give the name

of *blende* to a mineral which is chiefly the protosulphuret of zinc: its colour is yellowish, brown, or black almost like galena: its specific gravity is usually 3.9 or 4.

Protosulphuret. The above artificial compound, or the mineral, may be taken as examples of the union of 1 atom zinc and 1 sulphur. But the most correct and convenient way of forming it for the purpose of chemical investiagtion is, to drop a given portion of some salt of zinc into a dilute hydrosulphuret. A white precipitate falls, which when dried becomes a dark cream colour. It is found to consist of 2 parts zinc and 1 of sulphur nearly; that is, of 29 parts zinc and 14 sulphur.

Deutosulphuret, trisulphuret, &c. of zinc. These combinations may be made, up to the 5th or *quinsulphuret,* in the humid way by quadrisulphuret of lime, &c. The oxide may be first precipitated by lime water, or not, as we please, and then treated with quadrisulphuret according to the degree of sulphuration required. I found 100 measures of 1.29 nitrate of zinc with 2500 of 1.026 sulphuret of lime yield 40 grs. dry sulphuret zinc, of a yellowish white colour; the liquid was found to retain 13 or 14 grains of sulphur, by converting it into sulphuric acid by means of

oxymuriate of lime. The nitrate contained
11½ zinc and 2.8 oxygen; so that about 28
sulphur had combined with the zinc, and
about 14 remained in solution, or $\frac{1}{3}$ of the
whole, as has been already explained. By
proportion, if 11½ : 28 :: 29 : 70; or 1 atom
of zinc (29) combines with 5 atoms of sulphur (70). The intermediate combinations
I have not particularly examined; they do
not differ much in appearance from the one
just described; they burn blue and are reduced by it to the protosulphuret; and they
give sulphuretted hydrogen by muriatic acid.

27 and 28. Sulphurets of potassium
and sodium.

According to Davy and Gay Lussac, potassium and sodium unite with sulphur by
heat with vivid combustion. The compounds
appear to be protosulphurets, that of potash
being nearly as 35 potassium to 14 sulphur,
and that of sodium as 21 sodium to 14 sulphur. When potassium and sodium are heated along with sulphuretted hydrogen, an union likewise takes place; two atoms of gas
unite to one of the metals, except that 1 atom
of hydrogen is liberated, corresponding of
course in quantity to that liberated by treating

them with water. When the compound thus formed is treated with muriatic or sulphuric acid, the same quantity of sulphuretted hydrogen nearly is liberated that was originally combined. So that the compound may be regarded as sulphuretted hydrogen united to the protosulphurets. The colour of these sulphurets varies from grey to yellowor reddish.

29. Sulphurets of bismuth.

Protosulphuret. Bismuth combines with sulphur by heat, in the manner already described in the account of tin sulphurets. I found 100 parts bismuth in this way combine with 22 sulphur after 4 operations : this is therefore the protosulphuret or 1 atom bismuth (62) with 1 of sulphur (14). It may also be formed by substituting the oxide of bismuth for the metal. It has a dark brown or black metallic appearance, much like that of tin. It yields sulphuretted hydrogen in heated muriatic acid.

Hydrosulphuret of bismuth. When a solution of bismuth in nitro-muriatic acid is dropped into hydrosulphuret of lime, a black powder precipitates, which, when dried in the common temperature, appears to be hydrosulphuret of bismuth, or one atom sulphuretted hydrogen and one oxide of bismuth. It yields sulphuretted hydrogen by cold muriatic acid. But if the precipitate be dried in a heat of about 200°, the atom of water seems to be expelled, and there remains only the protosulphuret. Thus I found 69 parts oxide of bismuth unite to 15 sulphuretted hydrogen to form 84 hydrosulphuret of bismuth, when dried in the air; but upon being heated a little, it lost 8 parts of water and was reduced to the protosulphuret, retaining in great part the same appearance as before.

Deutosulphuret and trisulphuret of bismuth with oxygen. When nitro-muriate of bismuth is thrown into water the oxide is precipitated; if the acid water be decanted, quadrisulphuret of lime be put to the moist oxide and due agitation be used, the oxide abstracts sulphur from the lime so as to obtain 2 or 3 atoms for each one, if the sulphur be sufficient in quantity. To 6 oz. water I put 100 grain measures of 1.286 nitro-muriate, which I knew from its formation contained 20 oxide; after the precipitate had subsided I poured off 5 oz. of acid water, and to the remaining precipitate diluted with water I put 300 of 1.056 sulphuret of lime and agitated for 10 minutes. There were obtained 33 grains of brownish black sulphuret of bis-

muth dried for some hours in a temperature of
120°. I put the above 33 grains into a gas bottle with 100 muriatic acid and boiled it; there
were obtained only 2 or 3 cubic inches of sulphuretted hydrogen, the oxide was dissolved
and sulphur liberated; the sulphur collected
and dried weighed 9 grains, and the oxide
precipitated again from the muriatic acid by
water and dried, weighed 17 grains, besides
loss. From this it is evident the oxygen of
the oxide must have been chiefly retained in
the compound, and must have united to 2,
and in great part to 3, atoms of sulphur. For
20 oxide would require 12 sulphur to form
trisulphuretted oxide; and there was evidence
of its having nearly, if not wholly, that
quantity.

30. *Sulphurets of antimony.*

Protosulphuret. This is a natural production, and found in the state of a dark grey
mineral of metallic appearance, and of the
sp. gr. 4.2. It may also be formed artificially
by uniting metallic antimony and sulphur by
heat. Most authors nearly concur in assigning to it 74 parts antimony and 26 sulphur,
per cent. That is, 1 atom antimony (40) and
and 1 of sulphur (14). It yields sulphuretted

hydrogen by muriatic acid and heat, and a
solution of the metallic oxide is obtained.

Hydrosulphuret. When antimony is precipitated from a solution, by sulphuretted hydrogen or a hydrosulphuret, or from an alkaline solution of the sulphuret by an acid, it appears in the form of an orange yellow powder, denominated golden sulphuret. It is
constituted of 1 atom sulphuretted hydrogen
and 1 of protoxide of antimony; it readily
yields sulphuretted hydrogen by muriatic acid,
and the oxide combines with this acid. Exposed to heat, water is expelled and protosulphuret left. It is constituted of 40 antimony,
7 oxygen, 14 sulphur and 1 hydrogen; or of
54 protosulphuret and 8 water.

Bisulphuretted, trisulphuretted and *quadri-
sulphuretted oxide of antimony.* When crystallized muriate of antimony is agitated along
with dilute quadrisulphuret of lime, an orange
yellow compound is formed, consisting of the
oxide and sulphur. To 350 quadrisulphuret of
lime, diluted with lime water, I put 22 grains
moist crystals of muriate, and agitated well
for some time. Got 26 grains dry yellow sulphuret, which heated burned blue, and left
from 13 to 14 black grey sulphuret, equal to
10 antimony nearly; hence it must have been
a quadrisulphuret, or rather sulphuretted ox-

ide; for, by heating this compound in muriatic acid, a solution is obtained and sulphur liberated without the extrication of gas. Less of the sulphuret of lime would have produced a sulphuret of the same colour, but containing less of sulphur; so that it is evident various proportions may exist in combination. Instead of the crystallized muriate, the recently precipitated oxide, nearly free from acid, may be used to produce these compounds.

31. *Sulphuret of tellurium.*

Tellurium unites with nearly its weight of sulphur, by heat, according to Davy. It is probable that as usual in such cases, a protosulphuret is formed. This would lead to the conclusion that the atom of tellurium is only equal in weight to that of sulphur; which does not accord with results from the other combinations of tellurium, and hence the above fact may not perhaps be sufficiently ascertained.

32. *Sulphurets of arsenic.*

Arsenic may be combined with sulphur by exposing a mixture of the metal and sulphur or of the white oxide and sulphur, to a heat approaching to redness. In the latter case more sulphur is required, because the oxygen is carried off in the shape of sulphurous acid. Three parts of arsenic with two, three or more of sulphur may be used; the heat should be less if a greater proportion of sulphur is intended to be united. As both the elements are volatile in a moderate heat, and that in unequal degrees, considerable difficulty occurs in ascertaining by the synthetic mode, the proportions of the elements combined; if too little heat be used, only a mechanical mixture is obtained, of any proportions we please; if too much heat be used, part of the arsenic as well as part of the sulphur sublimes, and the sulphuret itself sublimes at a heat not much exceeding that required for their union. Hence in a great measure we have the discordant results of those who have taken the synthetic method. The analytic method is to be preferred, and those who have taken it have succeeded the best; but even this is attended with greater difficulties than with most of the other sulphurets.

The artificial sulphurets of arsenic constitute two varieties chiefly, and these are also found native in various parts of the earth.

1. *Protosulphuret.* Native sulphuret of arsenic, called orpiment, is found in Turkey

and elsewhere in considerable masses; when
broken it exhibits a foliated structure, somewhat flexible, and of a brilliant golden yellow colour. Its specific gravity is usually
about 3.2; at least that was the case with the
specimen I used. When heated so as to be
near melting, its surface reddens, probably
by the loss of sulphur. The same sulphuret
is procured artificially in the humid way
whenever a solution of the oxide of arsenic
in water, &c. is treated with sulphuretted hydrogen, or a hydrosulphuret, and afterwards
with an acid; or when this or any other species of sulphuret of arsenic is dissolved in an
alkali and the solution treated with an acid.
Kirwan in 1796 states, that it is generally
thought to consist of 100 arsenic and 11 sulphur, but that Westrumb says it contains 100
arsenic and 400 sulphur, which Kirwan thinks
more probable; they are both however very
wide of the truth. Thenard, in the 59 Vol.
of the An. de Chimie, 1806, asserts that it
consists of 100 arsenic and 75 sulphur; but
he does not point out the experiments on which
this result rests; and it is not very near the
truth. Laugier in the same An. Vol. 85, for
1813, in a paper of great merit, finds the native
orpiment to contain 38 per cent. of sulphur;
his method is to dissolve the orpiment in warm

dilute nitric acid; to precipitate the sulphuric acid by nitrate of barytes, and from the
sulphate of barytes infer the sulphur; the rest
he considers as arsenic, not knowing how to
detach the arsenic acid from the nitric acid so
as to determine the arsenic by experiment. I
have pursued this method with the advantage
of being able to determine the arsenic as well
as the sulphur: Ten grains of orpiment in
fine powder were dissolved in 100 measures
of 1.346 nitric acid diluted with as much water, by digesting in a heat so as to keep a
constant moderate effervescence for about 2
hours. The liquid obtained, being diluted,
yielded 536 measures of 1.061. By carefully
and gradually dropping in muriate of barytes
I found 150 measures of 1.162 just sufficient
to saturate the sulphuric acid, and the sulphate of barytes produced dry was 28 grains,
the loss I estimated 1 grain: now one third
part being sulphuric acid, and $\frac{2}{5}$ of the acid
being sulphur, we have $\frac{2}{15}$ of 29 = 3.87, or
3.9 for sulphur. The residuary liquid was
then treated with lime water till an excess
was manifest, and produced no farther precipitate; the arseniate of lime was collected
and dried, and gave 16 grains. Now I had
determined by experiments hereafter to be related, that $\frac{4}{7}$ of arseniate of lime are acid

and $\frac{2}{3}$ of the acid are arsenic; hence $\frac{8}{27}$ of 16 = 6.1 for the arsenic, which added to 3.9 sulphur, make up the 10 grains of orpiment.

When this orpiment is treated with caustic alkali, it is completely dissolved; it is thrown down by acids I find unaltered. If 61 arsenic combine with 39 sulphur, 100 must take 64 nearly; which corresponds with 1 atom of each, or 21 arsenic + 13 or 14 sulphur.

Subprotosulphuret. Sulphur and arsenic are found native in certain places, combined in masses of a brownish red or orange colour and glassy fracture: this combination is called *realgar,* and is also manufactured in large quantities in Saxony, chiefly for the use of calico-printers. Its constitution and specific gravity vary considerably, owing chiefly I imagine to the greater or less heat to which it is exposed, and to the proportions of the elements in the first mixture. I have specimens of 3.3 and 3.7 sp. gr. ; and it is probable these are not the extremes; the heaviest is the darkest colour. Of course the heaviest contains the most arsenic, and I have reason to believe that the sp. gr. is nearly as good a test of the proportions of the elements as chemical analysis. Realgar when pulverized is of an orange colour: it is much sooner dis

solved in dilute nitric acid and requires less,
than the same weight of orpiment. Caustic
alkali dissolves it partially, taking up the
protosulphuret and leaving the excess of
arsenic, the quantity of which may hence be
ascertained. Ten grains of realgar took 80
measures of 1.347 nitric acid, diluted with
as much water; digested in a heat of about
150° it was all dissolved in $1\frac{1}{2}$ hour, and
yielded 536 liquid of 1.05 sp. gravity. This
treated as before gave 24 sulphate of barytes
= 3.2 sulphur, and 18 arseniate of lime = 6.9
arsenic. This result agrees nearly with Laugier's in regard to the sulphur in native realgar: but the artificial realgar, which he
made by combining arsenic and sulphur,
yielded him 40 per cent. sulphur by my estimation and 42 by his own: the sp. gravity
of his artificial realgar is not given. Westrumb estimates realgar at 100 arsenic and
25 sulphur, and Thenard at 100 arsenic and
33 sulphur. But from the above it must be
concluded to contain 100 arsenic and 45 to 50
of sulphur. One hundred parts of the same
realgar heated in caustic potash were resolved
into 78 orpiment taken up by the liquid and
22 arsenic precipitated.

It appears to me most probable that a true
subsulphuret would be most convenient for

the printers' use, or one containing 100 arsenic and 32 sulphur, that is, 2 atoms arsenic and 1 sulphur. The object being to deoxidize indigo and obtain it in solution in a green state, we may suppose that 1 atom arsenic takes the oxygen from the indigo and then forms arseniate of lime which precipitates, whilst the other atom in union with the sulphur, takes the green indigo and unites it to the potash, making a quadruple compound of arsenic, sulphur, green indigo and potash in solution. If this view be right the heaviest and darkest coloured r algar of commerce must be the most advantageous for this purpose. Some printers however prefer the protosulphuret.

Deutosulphuret. Proust, by heating 100 arsenic with 300 sulphur in one instance got 222 parts, and in another 234 parts of a transparent deep greenish yellow sulphuret, (Jour. de Phys. 59—p. 406. 1804). Now it is very remarkable that if we take the atom of sulphur at 13 and that of arsenic 21, one of this and two of the former will be found as 100 to 124, together 224; but if sulphur be 14, then the proportion will be 100 to 133, together 233. It seems more than probable that Proust had accidentally used that degree of heat in the combination which is requisite

for forming the deutosulphuret. It is probable too that Laugier always used a higher
heat, as he uniformly obtained the same
(lower) sulphuret whatever were the proportions, the excess of either being sublimed or
separated by the heat.

Trisulphuret, quadrisulphuret, &c. When
a solution of the oxide of arsenic is treated
with quadrisulphuret of lime, little precipitate appears; but if muriatic acid be dropped
in, a fine yellow precipitate is formed. This
I have reason to think is sometimes a trisulphuret, and at other times a quadrisulphuret
or higher; but it is difficult to investigate
these compounds, and on that account I speak
with some uncertainty.

33. *Sulphuret of cobalt.*

Sulphuretted hydrogen does not precipitate
cobalt from solutions containing that metal;
but hydrosulphurets precipitate it.

Protosulphuret. This compound is obtained
whenever a neutral solution of cobalt is treated with hydrosulphuret of lime, &c. or it may
be obtained from any acid solution by first
precipitating the blue oxide by an alkali, and
then introducing sulphuretted hydrogen into
the mixture. By this last method I found a

solution previously known to contain 44 parts by weight of protoxide to absorb 15 parts of sulphuretted hydrogen; when filtered and dried in a heat of 100° it yielded 51 parts of protosulphure.. In appearance it resembles many of the other black sulphurets. It consists of 100 cobalt and 38 sulphur; Proust finds 40 sulphur, but he considers it only an approximation.

The same sulphuret may be formed by heating the oxides of cobalt and sulphur together to a red heat; at least a combination is effected as Proust observed, but I have not investigated the proportions. Sulphur does not seem to combine with the metal in this way.

Deutosulphuret....dodecasulphuret. When the recently precipitated and moist oxide of cobalt, the neutral muriate, or acid muriate of cobalt, as well as other salts of the same, are treated with dilute quadrisulphuret of lime, sulphurets of cobalt are formed in various proportions according to the ingredients, from the deutosulphuret to the dodecasulphuret: these precipitates are all black and not easily distinguished in appearance; but there is reason to believe they are true chemical compounds.

34. *Sulphurets of manganese.*

Though sulphur and manganese do not unite directly, they can be brought into union by intermediate bodies, both in the dry and humid way.

Protosulphuret. This compound may be formed by heating to a low red, a mixture of the oxide of manganese and sulphur, or of the white carbonate of manganese and sulphur; or it may be formed by treating a solution of manganese by a hydrosulphuret, (sulphuretted hydrogen not producing any precipitate); this last method seems to produce a dry hydrosulphuret of manganese, which being heated to red nearly, parts with water and a little sulphur and there remains the protosulphuret. The protosulphuret is of a snuff brown colour; but the hydrosulphuret, when recently precipitated is of a light drab colour, which grows deeper when exposed to the air, and when dried becomes brown like the protosulphuret; when heated, the colour is not much changed. The hydrosulphuret of manganese gives sulphuretted hydrogen by cold muriatic acid, and the protosulphuret gives the same by the acid heated.

The proportion of the elements in the protosulphuret may be inferred from the fact that the black oxide yields its own weight of protosulphuret; that is, 156 grains, composed of 100 metal and 56 oxygen give 156 of sulphuret; hence the atom of metal, 25, unites with one of sulphur, 14. I found 32 of the protoxide in solution unite to 15 of sulphuretted hydrogen to form 47 hydrosulphuret dried in 100°. This lost about 8 parts or rather upwards by heat.

Deutosulphuret, trisulphuret and *quadrisulphuret.* These may be formed by treating neutral solutions of manganese, or the recently precipitated oxide, by quadrisulphuret of lime. They are formed somewhat slowly and by considerable agitation with a smaller or greater proportion of the lime sulphuret. They are all light drab, and are reduced to the protosulphuret by heat.

35. *Sulphuret of chromium.*

I have not had an opportunity of ascertaining whether chromium or its oxides combine with sulphur or not, though several attempts were made for that purpose.

36. *Sulphuret of uranium*

From the experiments of Bucholz it would seem that uranium may be combined with sulphur, but the proportions have not been ascertained. (An. de Chimie. 56—142.)

37. *Sulphuret of molybdenum.*

From Bucholz and Klaproth's analyses of molybdena it would seem that the native sulphuret consists of 60 metal and 40 sulphur; but it does not appear whether this should be considered as the protosulphuret or the deutosulphuret. If it is the protosulphuret the atom of molybdenum weighs 21, but if the deutosulphuret, the atom of metal weighs 42; and the atom of the sulphuret or molybdena must weigh either 35 or 70.

38. *Sulphuret of tungsten.*

According to Berzelius, a sulphuret of tungsten may be obtained, by heating a mixture of tungstic acid and sulphuret of mercury in the proportion of 1 to 4, in a crucible. The mixture in his experiment was covered with charcoal and the crucible inclosed in another containing charcoal; the whole was then exposed to the heat of a furnace for half an hour. The sulphuret obtained was a greyish black powder; it was found to consist of 100 metal and 33¼ sulphur, or about 3 metal to 1 sulphur. Hence this must be the deutosulphuret if we consider the atom of tungsten to be 84; but considering the high degree of heat to which it was exposed, it would seem more likely to be the protosulphuret; if so, the atom of tungsten must be considered as 42 only, or half of the other number.

39. *Sulphuret of titanium.*

No compound of titanium and sulphur has been formed.

40. *Sulphuret of columbium.*

This combination is unknown.

41. *Sulphuret of cerium.*

This combination is also unknown.

SECTION 15.

EARTHY, ALKALINE, METALLIC AND OTHER PHOSPHURETS.

Phosphorus like sulphur is capable of being combined with several of the earths and metals as well as with other bodies; but the combination is not so easily effected, and the products are less interesting than those of sulphur: from considerations of these circumstances together with those of the expence and danger in making experiments on phosphorus we may account for, this class of bodies being as yet imperfectly known.

Margraf in 1740 attempted to combine phosphorus with many of the metals; but his experiments were mostly unsuccessful.

Gengembre in 1783 endeavoured to unite phosphorus with the alkalies; in this he failed of success, but discovered the phosphuret of hydrogen, or the spontaneously inflammable gas now denominated phosphuretted hydrogen. (Journal de Physique, 1785.)

In 1786 Mr. Kirwan published some experiments on phosphuretted hydrogen, (Philos.

Trans.); he ascertained that water impregnated with this gas had the property of precipitating various metals from their solutions.

The ingenious and indefatigable Pelletier has more merit than any other person in his investigations of the phosphurets. An important memoir of his on the manufacture of phosphorus in the large, is given in the Journal de Physique for 1785; in this he states that 4 or 5 lbs. sulphuric acid are commonly requisite for 6 lbs. calcined bones; and that from 18 lbs. calcined bones he obtained by the usual process, 12 oz. of phosphorus. In 1788 he read an essay on the phosphurets of gold, platina, silver, copper, iron, lead and tin. (An. de Chimie, 1—106). In 1790 he published an essay on the combinations of phosphorus with sulphur. (*Ibid.* 4—1). An additional memoir was published in 1792 on the same metallic phosphurets; and another on the phosphurets of mercury, zinc, bismuth, antimony, cobalt, nickel, manganese, arsenic and the other metals.

M. Raymond in the An. de Chimie, 1791, recommends, instead of potash, moist hydrate of lime and phosphorus in order to obtain phosphuretted hydrogen with greater facility; and in the same Annals for 1800 he asserts

that water absorbs a considerable portion of
phosphuretted hydrogen, and becomes capable of precipitating metals from their solutions in acids, and of forming phosphurets,
in this respect resembling sulphuretted hydrogen.

Mr. Tennant discovered in 1791 that carbonic acid combined with the earths and alkalies is capable of decomposition by phosphorus, in a red heat; and Dr. Pearson, following up the discovery, found that pure or caustic lime may be united to phosphorus by heat
so as to form phosphuret of lime; and that
this dry compound when put into water is decomposed and gives out bubbles of phosphuretted hydrogen gas, which as usual explode
spontaneously on reaching the surface of the
water and coming into contact with the air.

In 1810 I published the method of analysing phosphuretted hydrogen by Volta's eudiometer; having found that this gas and oxygen may be mixed together in a narrow tube
without explosion and afterwards exploded as
other similar mixtures by an electric spark.

Dr. Thomson published an essay on phosphuretted hydrogen in the Annals of Philosophy for August, 1816. He agrees with me
very nearly as to the constitution and properties of this gas, as far as I have gone; but

he has ascertained several additional properties of the gas, which I shall advert to in the sequel.

Sir H. Davy and Gay Lussac have investigated several compounds of phosphorus, particularly with muriatic and oxymuriatic acids, and with the new metals potassium and sodium, which I shall have to notice in their proper places.

Other authors have written on phosphurets besides those I have mentioned, but they do not require to be particularly distinguished in this enumeration. We shall therefore proceed to describe the phosphurets more particularly.

1. *Phosphuret of hydrogen.*

From recent experiments which I have made on phosphuretted hydrogen gas, I find the account already given (VOL. 1. page 456) is deficient, and in several respects inaccurate; I shall therefore substitute the following, as more perfect and correct.

Phosphuretted hydrogen may be obtained nearly pure, by the methods recommended by Dr. Thomson. Phosphuret of lime that has been carefully secluded from the atmosphere, may be put into a small phial filled

with water, acidulated by muriatic acid; into
this a cork with a bent tube must be immediately put under water, so that the phial and
tube are both full of water; gas soon begins
to appear, which rising to the top of the phial,
expels a corresponding portion of water, and
in due time the gas itself comes over and may
be received as usual: if the phial in which the
gas is generated be warmed to 140 or 150°,
the gas is given out more readily. A half
ounce phial with 20 grains of phosphuret in
small lumps, will produce 3 or 4 cubic inches
of gas. If the phosphuret of lime has been
previously exposed for a few hours to the atmosphere, the gas is more abundant, but consists chiefly of hydrogen, mixed with a little
phosphuretted hydrogen.

Pure phosphuretted hydrogen is distinguished by the following properties : 1. It ex-
plodes when coming into the atmosphere in
bubbles, and a white ring of smoke subsequently ascends: 2. It is unfit for respiration,
and for supporting combustion: 3. Its specific gravity is 1.1 nearly, that of atmospheric air being unity: 4. Water absorbs fully $\frac{1}{8}$
of its bulk of this gas, which is expelled
again by ebullition or by agitation with other
gases, but not without some loss: 5. A small
portion being electrified for some time, de-

posits abundance of phosphorus, and expands
from one volume to $1\frac{1}{2}$ nearly, which is found
to be pure hydrogen: 6. Liquid oxymuriate
of lime absorbs phosphuretted hydrogen, converting it into phosphoric acid and water,
and leaves any free hydrogen that may be
present; hence we are enabled to ascertain
the proportion of free hydrogen in any such
mixture, an important point as far as regards
this gas: 7. One volume of pure phosphuretted hydrogen, requires two volumes of
oxygen for its complete combustion by an
electric spark, in Volta's eudiometer; (the
gases must be previously mixed in a tube not
more than $\frac{3}{10}$ of an inch in diameter, to prevent an explosion in the act of mixing, after
which they may safely be transferred into any
other vessel); the result of the combustion
is phosphoric acid and water: 8. One volume
of phosphuretted hydrogen, mixed with from
2 to 6 volumes of nitrous gas, may be exploded by electricity in Volta's eudiometer;
or it may be exploded by sending up a bub-
ble of oxygen, without electricity; in like
manner, may the mixtures of phosphuretted
hydrogen and oxygen be exploded by a bub-
ble of nitrous gas: 9. One volume of phosphuretted hydrogen, mixed with 4, less or
more, of nitrous oxide, is also explosive by

electricity, but the mixture undergoes no change without electricity, at least in a day: 10 Mixtures of phosphuretted hydrogen and nitrous gas have a slow chemical action, by which in from 1 to 12 hours, the phosphuretted hydrogen is burnt and the nitrous gas decomposed into nitrous oxide and azotic gas: 11. According to Sir H Davy and Dr. Thomson, phosphuretted hydrogen gas being heated along with sulphur in a dry tube, the gas is decomposed and a new gas, sulphuretted hydrogen, is formed, and the phosphorus unites with the sulphur. Davy says the gas is doubled in volume by this operation; but Thomson says it remains the same; some doubt therefore exists respecting this fact: 12. When phosphuretted hydrogen gas is let up to oxymuriatic acid gas, a quick combustion with a yellow flame is observed, and the result varies according to the proportions: when one volume phosphuretted hydrogen is put to 3 or 4 of acid gas, both of the gases disappear, and muriatic and phosphoric acids are produced.

As these properties differ in many respects from those hitherto assigned to this gas, it will be necessary to enlarge upon them. The sp. gr. of this gas has already been adverted to, (Vol. 1.), and its great variation from .3 to .85; more recently Dr. Thomson finds

it about .9. In all these instances it was, I
have no doubt, contaminated with less or
more of hydrogen; at least it was so in my
own instance; for, I have the proportion of
oxygen which it required for its complete
combustion, both before and after it was
weighed. It was what I then thought pure
gas: that is, 100 volumes required nearly 150
of oxygen; but I am now convinced that gas
of this description contains $\frac{1}{3}$ of its volume
of free hydrogen; hence the correction of the
sp. gravity. Davy estimates the sp. gr. of the
gas which he denominates *hydrophosphoric* at
.87 or 12 times that of hydrogen; this gas, as
will appear from this and other properties, is
in all probability phosphuretted hydrogen gas,
nearly pure.

The absorption of this gas by water, has
been stated variously. In 1799 M. Raymond
found that water absorbs rather less than $\frac{1}{4}$ of
its volume of this gas: in 1802, Dr Henry
rates its absorption at $\frac{1}{47}$ only; in 1810 I
found it $\frac{1}{27}$; in 1812, Davy found it (hydrophosphoric gas) to be $\frac{1}{8}$; in 1816, Dr. Thomson found it to be $\frac{1}{47}$; I now estimate it as
stated above at $\frac{1}{8}$. These enormous differences may be partly accounted for by varieties
in the gas; and partly from the theory of the
absorption not being understood; but these

are scarcely sufficient excuses in all the cases.
I find that my early experiments on the absorption of phosphuretted hydrogen by water,
were made prior to the discovery of the method of analysing the gas by electric combustion; consequently tney were deficient in regard to the quality of the gas, both before
and after agitation; the best gas that ever I
had, was such as took 150 oxygen per cent.
for its combustion, exclusive of any common
air; and it was often such as to require considerably less. The bottle which I used for the
purpose in 1810 contains 2700 grains of water; at first I charged water with hydrogen;
into this 120 grain measures of phosphuretted
hydrogen were put, and the whole well agitated: there were left 98 measures;—this
proved that the gas was more absorbable than
hydrogen: into the same water were put 98
more phosphuretted hydrogen and agitated;
out, 80; this confirmed the proof: Into the
same water were put 97 hydrogen and agitated
well; out 105: This shewed that the hydrogen
had expelled a part of the gas again, and was
less absorbable of the two. As the phenomena were much the same as if oxygen had
been used instead of phosphuretted hydrogen,
it was concluded to have the same absorbability.

In the present instance, however, I have
been more circumstantial; after repeatedly
agitating water with pure azotic gas, so as to
saturate it and expel the oxygen, I then put
in 110 grain measures of phosphuretted hydrogen composed of 100 pure gas, 5 hydrogen, and 5 azotic gas or rather atmospheric
air. After due agitation, all was absorbed
but 35; this was mixed with a known portion
of oxygen and exploded; the diminution was
19 measures; the oxygen remaining was determined by hydrogen; from which it appeared that 10 combustible gas had taken 9 oxygen. Now 10 being $\frac{2}{7}$ of 35, we may consider the water as $\frac{2}{7}$ impregnated with the
phosphuretted hydrogen, and $\frac{5}{7}$ with azote;
but as there were 105 combustible gas and
only 10 left, 95 must have entered the water
and caused it to be $\frac{2}{7}$ charged with the gas;
whence we may infer that 332 gas would
have been a full charge for 2700 water, which
is almost exactly $\frac{1}{8}$, as stated above. Other
experiments gave corresponding results. On
admitting 51 azotic gas to the water, and agitating it a good deal for 4 or 5 minutes, there
came out 51 measures or the same volume:
this was found in the same way to consist of
43 azote and 8 combustible, which took 10
oxygen. Again 51 azote was agitated in the

water, and there came out 51, of which 5 + were combustible and took 9 oxygen. After this the bottle of water was put into a pan of water which was raised to the boiling heat, a bent tube filled with water being adapted to the water bottle, and having its end immersed in water: by this operation gas was expelled from the water, and caught in the neck of the bottle; when it amounted to 22 grain measures it was transferred and was found to consist of 17 azote +5 combustible, which took 10 oxygen. By these experiments we see that the gas is expelled again from the water, both by ebullition and by other gases, nearly the same in quality, but much diminished in quantity, the reason of which is not very obvious. The liquid now required 30 measures of oxymuriate of lime, equivalent to 100 measures of oxygen, before it was saturated; that is, there appeared to be 50 phosphuretted hydrogen remaining in the water. Adding a little lime-water threw down a very sensible quantity of phosphate of lime.

The expansion of phosphuretted hydrogen by electricity is a subject on which there has been as much diversity as on its absorption. In 1797, Dr. Henry found that it expanded "equally with carbonated hydrogen." (Philos. Trans). In 1800, Davy states that phosphuretted hydrogen was not altered in volume by electricity. (Researches, page 303.) In 1810, my experiments led me to adopt the same conclusion. In 1811, Gay Lussac found (Recherches, page 214), that potassium heated in phosphuretted hydrogen gas, expanded 100 volumes to 146; he infers that the true expansion ought to have been to 150. In 1812, Davy observes, that when electric sparks are passed through gases of this kind, "usually there is no change of volume." (Elements of Chem. Philos. p. 294.) But he adds that when a gas (sp. gr. 6, hyd. being 1) was heated with zinc filings over mercury, there was an expansion of volume of more than $\frac{1}{7}$. Also potassium heated in it, made 2 parts become 3 or 3, parts rather more than 4, (1810); the residual gas in these cases was pure hydrogen. Hydrophosphoric gas (sp. gr. 12) yielded 2 volumes of hydrogen, by heating potassium in it. In 1816, Dr. Thomson found that by electric sparks phosphorus was deposited, and hydrogen remained "exactly equal to the original bulk of the phosphuretted hydrogen." Lastly, in 1817, I found by two experiments, that by electrifying 30 grain measures of phosphuretted hydrogen in a tube over water, uninterruptedly for nearly 2 hours, I produced an expansion of $\frac{1}{5}$, or the gas became 36 measures; originally the gas contained $2\frac{1}{2}$ common air, and the rest was combustible so that 100 measures took 190 oxygen. By exploding the residue with oxygen, I found that $\frac{1}{15}$ or $\frac{1}{20}$ of the phosphuretted hydrogen still remained undecomposed. Taking these observations into consideration along with the fact, that 1 volume of the purest gas requires 2 of oxygen for its combustion, I conclude that the true expansion should be $\frac{1}{3}$, or 3 volumes of gas should become 4, and then it will be found that $\frac{1}{3}$ of the oxygen is joined to the hydrogen and $\frac{2}{3}$ to the phosphorus, which accords with what appears to me the only correct view of the constitution of phosphoric acid, namely, 2 atoms of oxygen to 1 of phosphorus.

The action of oxymuriatic acid, whether free or combined, on phosphuretted hydrogen, is curious and interesting; in both cases it effects a complete and instantaneous combustion of both phosphorus and hydrogen; when the acid is put to in the state of gas, it not only burns the phosphuretted hydrogen, but any free hydrogen that may be present; but this has a limit: if the phosphuretted hydrogen be largely diluted (90 per cent.) with hydrogen, this last is wholly left; the reason seems to be, the phosphuretted hydrogen burns

at a lower temperature; and hence probably
it is, that liquid oxymuriate of lime burns the
phosphuretted hydrogen, but not the hydrogen gas.

The quantity of oxygen necessary to saturate a given volume of phosphuretted hydrogen is easily found. Oxygen gas containing
a known per centage of azotic gas, must be
used in some excess, mixed with a due portion of the gas. After exploding the mixture, the loss must be observed, and then the
remaining oxygen must be found by exploding
it with hydrogen. Hence the true volume of
oxygen spent by the first explosion, and that
of the combustible gas are both determined.
The due proportion of oxygen is so nearly 2
to 1, that I have not been able to determine
on which side the truth lies. Dr. Thomson
says that when phosphuretted hydrogen and
oxygen are mixed, two volumes to one, a white
smoke takes place, the volume of oxygen
gradually disappears, and there remains be-
hind a quantity of hydrogen exactly equal to
the original volume of the phosphuretted hydrogen. I have observed nothing at all like
this. A mixture of phosphuretted hydrogen
and oxygen stood 24 hours without sensible
diminution, and afterwards being exploded,
2 volumes of oxygen disappeared for 1 of phos-

phuretted hydrogen, the same as would have
done at the moment of mixing. Perhaps the
temperature may have some influence; mine
was about 55°.

I have tried the minimum of oxygen that
will consume or dissipate phosphuretted hydrogen gas. It may be exploded with about
¼ of its volume of oxygen, with the same
phenomena as Davy observed of the hydrophosphoric gas. Phosphorus is thrown down
and a volume of combustible gas is left about
10 per cent. greater than the original volume
of phosphuretted hydrogen. This gas is nearly
pure hydrogen. Hence the whole gas may
be dissipated at 2 successive explosions, by
rather less than an equal volume of oxygen.
If phosphuretted hydrogen be exploded with
an equal volume of oxygen, phosphorous
acid, water and a little phosphoric acid are
formed, and some hydrogen remains.

One of the most remarkable properties of
phosphuretted hydrogen, is that announced
by Dr. Thomson, namely, its combustion with
nitrous gas by electricity; and the *slow* combustion by the same gas, which I have mentioned above is a fact still more difficult to explain. I tried the combustion of phosphuretted hydrogen by nitrous gas and electricity in 1810, but did not succeed. The rea-

son was, the gas was not sufficiently pure. No phosphuretted hydrogen that is not 70˙or 80 per cent. pure, can, I imagine, be exploded by nitrous gas; even the purest requires sometimes more than one spark, when mixed in the most favourable proportions; and I have known instances in which the mixture has exploded after electrification for a few minutes. An excess or defect of nitrous gas, occasions oxygen or hydrogen to be found in the residual gas, just as when we explode with oxygen. One volume of phosphuretted hydrogen requires, as nearly as I can find, $3\frac{1}{2}$ of nitrous gas for mutual saturation. The azote developed amounts to $1\frac{3}{4}$ volumes or rather less, (due allowances in all such cases being made for that already existing in the two gases.)

The mutual action of nitrous gas and phosphuretted hydrogen without electricity exhibits one of the most singular phenomena we have in chemistry. Nitrous gas seems constantly to be decomposed, one part producing nitrous oxide and another part azote, even though an excess of nitrous gas remain undecomposed in the mixture, and both the phosphorus and hydrogen are completely burnt; but if the nitrous gas be deficient, then nitrous oxide, azote, and some of the phosphuretted hydrogen are found in the residue, and the

rest of the phosphuretted hydrogen is completely burnt or converted into phosphoric
acid and water; here appears no preference
of phosphorus to hydrogen in this case, nor
any partial combustion. From an attentive
consideration of the results of several experiments, I am inclined to offer the following
solution of this remarkable case : One atom
of phosphuretted hydrogen attacks 5 of nitrous gas at the same instant; the atom of
phosphorus takes 2 of oxygen, and gives the
corresponding 2 of azote to the two of nitrous
gas, and thus makes two atoms of nitrous oxide, while the hydrogen takes 1 of oxygen from the fifth atom and liberates the
azote; thus 2 measures of nitrous oxide are
formed along with 1 of azote; and they are
generally found in the residue in that ratio.
The azote does not seem to pass through the
intermediate state of nitrous oxide; for, as
soon as the nitrous gas ceases to exist, there
is an end of the combustion.

It may be proper to advert more particularly to the hydrophosphoric gas of Davy.
That this gas is the same as that we have
been describing, can hardly admit of a doubt.
Their near agreement in sp. gr., in their absorbability by water, in the quantity of oxygen requisite for their combustion, in their

moderate expansion by burning with a minimum of oxygen and in their combustibility by oxymuriatic acid, are circumstances sufficient to warrant their identity. It is said that by heating potassium in this gas, one volume yields two of hydrogen; but it has not been found to yield two volumes by electricity, the more accurate criterion. Besides, both Davy and Gay Lussac find that potassium heated in the more common phosphuretted hydrogen expands it from 1 to $1\frac{1}{7}$ or $1\frac{1}{2}$ volume, which common electricity will not do; it is presumed therefore that the potassium in some way conduces to the production of a portion of the hydrogen. Spontaneous ignition or explosion is, I believe, no distinctive mark of variety in phosphuretted hydrogen; when this gas is *produced*, it is usually explosive from the uncombined phosphorus which it elevates; but the best and purest phosphuretted hydrogen loses the property wholly or partially by standing a while over water, though it loses no sensible part of its phosphorus.

It is commonly stated that phosphuretted hydrogen deposits phosphorus by long standing. This seems to be true; but the deposition is slower than I imagined. Seven years ago I set aside a bottle of impure phosphuretted hydrogen which I then labeled, 10 combustible take 14.6 oxygen; this bottle has not been preserved with special care to seclude the atmosphere; notwithstanding that, it is now such, that 10 combustible take 6.7 oxygen, and hence it still contains some genuine phosphuretted hydrogen.

2 and 3. *Phosphurets of carbone and sulphur.*

See Vol. 1. page 464.

4. *Phosphuret of lime.*

This compound may be formed by subliming phosphorus in a glass tube containing small fragments of recently calcined lime, heated to a low red. The sublimed phosphorus coming into contact with the hot lime, the two unite with a vivid glow, and in due time mutual saturation is produced. The result is a dry, hard compound of a deep brown or reddish colour, which on cooling must be put into a bottle and well corked, if not intended for immediate use, as it soon changes by the action of atmospheric air and moisture. With this precaution, I have reason to think it may be kept unimpaired for years.

As far as I know, no experiments have
been published relating to the proportion in
which phosphorus and lime unite. M. Dulong, in a valuable paper on the combinations of phosphorus and oxygen, in the
Memoires de la Societe d'Arcueil, VOL. 3.
(1817,) has given some account of his experiments on the earthy phosphurets; but
it is to be regretted that he has given none on
the proportions of their elements.

In order to ascertain the phosphorus, I put
10 grains of well preserved phosphuret of lime,
into 1000 grains of liquid oxymuriate of lime,
such that by previous trials I knew would impart 3.5 grains of oxygen; to this mixture a
quantity of muriatic acid was put, sufficient
to engage the lime; the phosphuretted hydrogen disengaged, was of course made to pass
through the liquid as it was generated, and became oxidized, so as to lose its gaseous form;
the surplus gas was prevented from escaping
by an inclination of the bottle; it was 45
grain measures only, and of this 30 were found
to be pure hydrogen, and the rest atmospheric air detached from the water; these 30
measures were the *free* hydrogen, which would
have been mixed with the phosphuretted hydrogen, in the ordinary way. In due time,
the whole of the phosphuret of lime was dis-

solved. The liquid was strongly acid, and manifested no smell of oxymuriatic acid, a proof that it was all decomposed. To this were added 70 more of the oxymuriate of lime before the smell of it was permanently developed. The liquid was next saturated with lime-water, and the phosphate of lime carefully collected and dried; when heated to a low red it weighed 12 grains, and consisted, according to my estimate of this compound, of 6— grains of phosphoric acid and 6 + grains of lime. The 6— grains of acid contained 2.4 phosphorus and 3.5 of oxygen. It must be remembered that 10 grains of phosphuret yield about 500 measures of phosphuretted hydrogen, and these contain 650 measures of hydrogen, which last is also oxidized at the expence of the oxymuriatic acid; but then there is an equivalent of oxygen from the water, so that this does not influence the calculation for oxygen. There appears then to be only an excess of .24 grains of oxygen unaccounted for, (arising from the additional 70 of oxymuriate of lime), which is as little as can be expected in such an experiment. If the phosphorus amount to 24 per cent. we may reasonably infer tha the remainder (76) is mostly lime, though I have not been able to detect above 60. Now if an atom of phos-

phorus weigh $9\frac{1}{3}$ and one of lime 24, the due
proportion of the protophosphuret of lime
would be 28 phosphorus and 72 lime; but
when the article is made for sale, it is more
likely to find a defect than an excess of phosphorus.

According to Dulong, when the earthy
phosphurets are decomposed by water, phosphuretted hydrogen and subphosphorous acid
are formed. I believe this determination is
right; for I find at most only $\frac{1}{3}$ of the above
proportion of phosphorus in the phosphuret-
ted hydrogen yielded by 10 grains of the
phosphuret of lime; the remaining $\frac{2}{3}$ seem to
rest in the liquid in combination with the oxygen and lime; that is, 1 atom of hydrogen
combines with 1 of phosphorus, and 1 of
oxygen with 2 of phosphorus. Notwithstanding this, the phosphoric acid produced
from the residue by means of oxymuriate of
lime, does not in general correspond to the
above quantity. Perhaps this loss may be
owing to the phosphorus carried over in mechanical suspension by the gas.

M. Dulong observes, that even the earthy
subphosphites are very soluble; this did not
appear to me to be the case with that of lime:
10 grains of phosphate of lime, that had
been exposed for 20 minutes to the air, were

put into a gas bottle filled with 400 grains of
water; this was kept at nearly the boiling
heat for an hour, when 725 grain measures of
gas were produced, and some phosphorus was
carried over with it into the receiving bottle and
bason of water. The gas being analysed, was
found to consist of 62 per cent. phosphuretted hydrogen, 33 hydrogen and 5 common
air. The 400 grains of water in the gas bottle treated with oxymuriate of lime, and then
with lime water, scarcely gave any appreciable quantity of phosphate of lime. The
insoluble residue when dried yielded 9 grains.
This dissolved in muriatic acid left a fraction
of a grain of dirty yellow powder, which indicated some phosphorus; and the muriate of
lime indicated about 6 grains of lime.

5. *Phosphuret of barytes.*

The combination of phosphorus and barytes may be effected in the same way as the
foregoing, and the compound has the same
appearance. According to Dulong, who has
examined this phosphuret with particular attention, it gives out phosphuretted hydrogen
when dropped into water, the same as that
of lime. When the gas ceases to be given

out, a powder remains completely insoluble in water, of a variable colour, yellow, grey or brown. It is not altered by the air; but it gives out a slight phosphoric flame when heated. Dilute nitric or muriatic acid, dissolves nearly the whole with a trace of phosphuretted hydrogen, and leaves only a few atoms of greenish yellow powder, soluble in oxymuriatic acid. The part dissolved by the acids being precipitated by ammonia, gives phosphate of barytes. From these facts he infers that the residue insoluble in water, consists of a small portion of phosphuret of barytes with excess of base, and phosphate of barytes. The water in which the phosphuret was decomposed, contains most of the barytes: carbonic acid produces a slight precipitate, and then leaves a neutral liquid containing the subphosphate of barytes, which appears to be a very soluble salt. Sulphuric acid throws down the barytes and leaves the subphosphorous acid in the liquid.

Nothing certain is determined from experiment respecting the proportion of phosphorus and barytes which combine; but from analogy it is probable that they combine atom to atom, or 68 parts barytes with 9 of phosphorus; or 100 parts of the compound contain 88 of barytes and 12 of phosphorus.

6. *Phosphuret of strontites.*

Phosphuret of strontites may be formed as the two preceding articles. It is in all respects similar to the phosphuret of barytes according to Dulong, and its properties therefore need not be particularized.

From analogy, I should apprehend, it must be constituted of 46 strontites and 9 phosphorus, or one atom of strontites to one of phosphorus; that is, 100 parts of phosphuret should contain 83 strontites and 17 phosphorus.

———

Combinations of the other earths and phosphorus have not yet been effected. Neither have the alkalies been combined with phosphorus· the hydrates of these as well as those of the earths, yield phosphuretted hydrogen when heated with phosphorus, and probably a phosphate or subphosphate of the base. M. Sementini of Rome is said to have succeeded in combining potash and phosphorus by means of alcohol. His experiments, however, appear to me too indefinite to warrant the conclusion. (See An. of Philos.—7. p. 280). The compounds of phosphorus with potassium and sodium are described in the sequel, amongst the metallic phosphurets.

7. *Phosphuret of gold.*

M. Pelletier heated together in a crucible, half an ounce of pure gold, one ounce of phosphoric glass and $\frac{1}{8}$ of an ounce of powdered charcoal, the heat was raised sufficiently to fuse the gold. Phosphoric fumes arose, but the whole of the phosphorus was not dissipated. The gold remaining was whiter than natural, and brittle under the hammer. Exposed to a very high heat it lost $\frac{1}{24}$ of its weight, and resumed the ordinary characters of gold.

The same chemist heated 100 grains of pure gold in filings to a bright red; he then projected small fragments of phosphorus amongst the gold successively till after it had entered into fusion. The gold preserved its colour, but became brittle under the hammer and granular in the fracture; it had increased 4 in weight.

Mr. Edmund Davy, by heating in a tube deprived of air, finely divided gold and phosphorus, effected a combination of them. It had a grey colour and metallic lustre. The heat of a spirit lamp was sufficient to decompose it. It contained about 14 per cent. of phosphorus. (Davy's Chemistry, page 448—An. 1812).

Oberkampf and Thomson have successively observed the precipitation occasioned by water impregnated with phosphuretted hydrogen, in solutions of muriate of gold. The former of these has some interesting remarks on the phenomena. When a current of this gas is passed through a dilute solution of muriate of gold for a time, and then suddenly discontinued, the solution becomes brown and passes soon to a fine deep purple. A yellowish brown precipitate is obtained, which is metallic gold, and the liquid, now become yellow again, contains muriate of gold and phosphoric acid. The experiment may be continued with the like results. But if the liquid be saturated with gas before any precipitate is suffered to subside, a black powder is obtained which does not seem to contain any metallic gold, and the liquor ceases to have any colour. This black powder is the phosphuret of gold; exposed to heat it inflames and leaves metallic gold, but its elements are not separable by mechanical means. (An. de Chimie, 80—146, for 1811).

Water impregnated with the gas was found to have like effects as the gas itself. Whence

Oberkampf concludes that as long as an excess of gold remains in solution, the phosphuretted hydrogen precipitates the metal only; but when the gas is in excess, the phosphorus leaves the hydrogen and unites with the precipitated gold.

I should rather suppose that the precipitation of the gold may be, in part at least, owing to the *free* hydrogen which we now know accompanies the phosphuretted hydrogen largely, in the manner in which this gas was formerly procured; however that may be, I find that water, impregnated with the purest phosphuretted hydrogen, has the property of precipitating the black phosphuret of gold from the muriate of that metal, in such manner as to effect complete mutual saturation, leaving nothing in the liquid but the muriatic acid. Let a solution containing a known quantity of gold be gradually dropped into water, containing a known quantity of phosphuretted hydrogen, as long as any black precipitate is formed. The point of saturation will be found when 60 parts by weight of gold have united to 9 of phosphorus, nearly; or when one atom of gold has united to one of phosphorus. Hence it may be concluded that 100 grains of the phosphuret of gold contain 13 or 14 of phosphorus, which agrees very nearly with the results of Mr. Edmund Davy abovementioned.

8. *Phosphuret of platina.*

M. Pelletier succeeded in combining platina with phosphorus by the same methods as with gold. By projecting phosphorus on grains of platina heated to a strong red, the latter acquired an increase of weight of 18 on the hundred; but this was probably an excess, as some vitreous phosphoric acid was found mixed with the mass.

In the Philos. Magazine, Vol. 40, Mr. E. Davy has related some experiments made with a view to combine platina and phosphorus; he effected it by heating platina and phosphorus together in an exhausted tube; the union commenced below a red heat and was attended with vivid ignition and flame. The compound was of a blueish grey colour and consisted of $82\frac{1}{2}$ platina and $17\frac{1}{2}$ phosphorus according to his estimate. Also by heating the ammonia-muriate of platina with $\frac{2}{3}$ of its weight of phosphorus in a retort over mercury, muriatic gas was liberated, and muriate of ammonia and phosphorus were sublimed, but there remained at dull red heat an iron black or dark grey mass at the bottom, of the sp. gr. 5.28. It was estimated to consist of 70 platina and 30 phosphorus; but I doubt whether it could consist of these two elements only.

Phosphuretted hydrogen water scarcely has any effect on muriate of platina. After some time a very light flocculent matter appears, as Dr. Thomson has observed; but this seems to me to be nothing but a slight precipitation of phosphorus alone; I apprehend the gas unites with the platina, but the compound remains in solution somewhat in the same manner as platina and sulphuretted hydrogen. The platina may be precipitated from the clear liquid by muriate of tin, much the same in appearance as if no phosphuretted hydrogen were present.

9. Phosphuret of silver.

When pieces of phosphorus are dropped amongst silver heated to red in a crucible, the two unite and enter into fusion, according to Pelletier; when the metal is saturated with phosphorus the whole continues in a state of tranquil fusion; but being withdrawn from the fire, at the moment of congelation, a

quantity of phosphorus becomes suddenly volatile and burns vividly, and the surface of
the metal becomes uneven. The metal on
being cooled, is found to have gained from 12
to 15 per cent. ; and he apprehends that when
fluid it contains 10 per cent. more, making
in all 25 phosphorus to 100 silver.

The phosphuret of silver is white and crystalline, brittle under the hammer, but capable of being cut with a knife. By a strong
heat the phosphorus is dissipated and leaves
the silver pure.

Both Raymond and Thomson observe that
phosphuretted hydrogen water precipitates
silver from its solutions of a black colour. I
find that a solution of sulphate of silver containing one grain of the metal, requires water containing 90 grain measures of phosphuretted hydrogen to saturate it; the whole of
the silver falls readily and leaves nothing but
the acid in the water. Now the weight of 90
measures of this gas is nearly $\frac{1}{9}$ of a grain;
hence the proportions of metal and phosphorus are as 10 to 1, which shows that they
combine atom to atom, or 90 silver to $9\frac{1}{3}$
phosphorus. This is somewhat less of phosphorus than is determined above by Pelletier.

10. *Phosphuret of mercury.*

M. Pelletier made several attempts to combine phosphorus and mercury. He seems to have succeeded best, by exposing mercury in an extreme state of division, to phosphorus under water in a moderate heat. The phosphuret is a black compound, which is resolved again into its elements by distillation.

When nitrate of mercury is treated with phosphuretted hydrogen water, a copious dark brown or black precipitate is instantly formed, as Raymond and Thomson have observed. This black precipitate, Raymond adds, soon becomes white and crystalline in passing from phosphuret to phosphate, by attracting oxygen.

I have found the black powder when dried in a moderate heat to abound in small shining globules, which have all the appearance of revived mercury. However this may be, I find that a certain weight of mercurial salt requires a certain portion of gas to saturate it, so as that the whole mercury shall be precipitated. One grain of mercury requires rather more than $\frac{1}{18}$ of its weight or 50 grain measures of the gas for its saturation. This proves

the combination to be the most simple, or atom
to atom; that is, 167 mercury take $9\frac{1}{3}$ phosphorus; or 100 mercury take $5\frac{1}{2}$ phosphorus
nearly.

11. *Phosphuret of palladium.*

When nitrate of palladium is dropped into
phosphuretted hydrogen water, a copious
black flocculent precipitate is immediately
formed, which doubtless consists of palladium and phosphorus.

Into 800 grains of phosphuretted hydrogen
water containing 20 grain measures of gas,
were put by degress 22 grain measures of
muriate of palladium (sp. gr. 1.01) containing .12 acid and .14 oxide, corresponding to
.12+ metal; mutual saturation was produced,
and a finely distinct black powder precipitated, leaving the water clear and colourless,
which was found by lime-water to contain .12
parts of a grain of muriatic acid. The black
powder collected and dried, corresponded as
nearly as could be determined in weight to
the ingredients. Now 20 measures of gas
would weigh .025 of a grain, of which .0025
would be hydrogen and .0225 phosphorus;
whence we have .12+ metal joined to .0225

phosphorus or 50 to 9 nearly, indicating one atom of each. Hence 100 palladium would take 18 or 19 phosphorus.

12. *Phosphuret of copper.*

M. Pelletier combined copper and phosphorus by the same means as the preceding compounds. One hundred grains of copper united by heat with 15 of phosphorus; the fused mass when cooled was white and very hard. As part of the copper gets oxidized during the process he thinks it probable, with M. Sage, that copper may acquire 20 per cent. of phosphorus.

In the 3d VOL. of Memoirs of the Society of Arcueil, page 432, M. Dulong converts fine copper wire into phosphuret by heating it to a low red, and passing the vapour of phosphorus over it in hydrogen gas. In the sequel he observes that 10 grammes of phosphuret of copper contained 1.97 of phosphorus; that is, the copper was to the phosphorus as 8.03 : 1.97, or as 100:24.5. This exceeds much Pelletier's result, and is, I think, too high. For, he found that the above phosphuret converted into phosphate of copper by nitric acid yielded 14.44 grammes. Now supposing the atom of phosphorus to weigh $9\frac{1}{3}$, that of phosphoric acid $23\frac{1}{3}$, and that of the black oxide of copper 70, we have an atom of phosphate of copper $= 93\frac{1}{3}$: and if $93\frac{1}{3} : 9\frac{1}{3} :: 14.44 : 1.444$, for the phosphorus in 10 grammes; and hence the copper would be 8.556: this would give 100 copper to 17 phosphorus nearly, which would accord well with Pelletier's determination, and very nearly agree with the theoretic result of 100 copper to $16\frac{2}{3}$ phosphorus.

Both Raymond and Thomson remark that phosphuretted hydrogen water produces a black or dark brown precipitate in sulphate of copper. I have not found any precipitate from any of the salts of copper by the same means. But if the blue hydrate be first precipitated by lime water, and then the phosphuretted hydrogen water admitted, the hydrate is immediately converted into a dark olive, which in all probability is a phosphuret of copper. From some experiments I am inclined to believe that this compound is the deutophosphuret, or two atoms of phosphorus to one of copper; and hence the copper is to the phosphorus as $100 : 33\frac{1}{3}$.

13. *Phosphuret of iron.*

M. Pelletier formed a phosphuret of iron by both the methods above described for gold. He describes the phosphuret as very hard, of a white colour, striated and magnetic. He estimates, with some uncertainty, that 100 iron may combine with 20 phosphorus.

Berzelius produced a phosphuret of iron by reducing the phosphate of the metal by charcoal and heat. (An. de Chimie, July 1816). He describes it as having the colour of iron, brittle and slightly acted upon by the magnet. By his analysis it was constituted of 100 iron and 30 phosphorus. The true proportion probably would be one atom to one, or 25 iron to $9\frac{1}{3}$ phosphorus; that is, 100 iron to 37 phosphorus.

Both Raymond and Thomson found that sulphate of iron yields no precipitate by phosphuretted hydrogen water; and I may add, that the precipitated oxide or hydrate is also unaffected by the same.

14. *Phosphuret of nickel.*

By projecting phosphorus amongst red hot nickel, Pelletier united 20 parts of the former to 100 of the latter. A part of the combined phosphorus, he observes, flies off on cooling, so that the above proportion may perhaps be too low. Theoretically one atom of nickel should combine with one of phosphorus; that is, 26 with $9\frac{1}{3}$, or 100 with 36.

I find that neither the nitrate of nickel nor the hydrate are affected by phosphuretted hydrogen water.

15. *Phosphuret of tin.*

Margraff was the first who combined phosphorus and tin by fusing the metal along with fusible salt from urine (phosphate of ammonia). Pelletier succeeded also in this way, as well as by the direct one of projecting phosphorus into melted tin. The compound was of a white colour; it gained 12 per cent. of weight; but as part of the tin was oxidized and adhered to the crucible in form of glass, he conjectures that tin would take from 15 to 20 per cent. of phosphorus. The atom of tin being 52, and that of phosphorus $9\frac{1}{3}$, the due proportion would be 100 tin to 18 phosphorus.

Phosphuretted hydrogen water does not seem to precipitate tin from solutions, nor yet to act upon the precipitated oxide.

16. *Phosphuret of lead.*

Lead combines with phosphorus by the same methods as tin ; but it is difficult to ascertain the proportions, according to Pelletier, from the oxidation and vitrification of part of the lead. Muriate of lead distilled with fusible salt of urine, also yielded phosphuret of lead. He conjectures the increase by phosphorus to be 12 or 15 per cent. ; but by theory it should only be 10 or 11 per cent.

Raymond says that the nitrate of lead is decomposed by phosphuretted hydrogen water, but with less force than salts of silver and mercury ; and that a phosphuret of lead is formed, of which he gives no character, except that it becomes in time a phosphate. Thomson says a slight white powder is formed by the mixture. This was the case with me ; but I suspected that the white powder was merely a little sulphate of lead, arisin from the impurity of the (rain) water ; and this was found to be the fact ; for the milkiness was just the same with like water unimpregnated with the gas. Besides, after the phosphuretted hydrogen water has been saturated with nitrate of lead till no more effect is produced still the water retains its peculiar smell, and a copious black precipitate is instantly produced by nitrate of silver or mercury. It appears then that phosphuret of lead cannot be formed this way. Neither does phosphuretted hydrogen water seem to have any effect on the recently precipitated oxide of lead.

17. *Phosphuret of zinc.*

Both zinc and its oxide seem to combine with phosphorus, according to Pelletier; but the proportions were not ascertained. By theory, zinc should take 32 per cent. of phosphorus.

18. *Phosphuret of potassium.*

Some account was given by Davy, of the combination of potassium and phosphorus in essays from 1807 to 1810; and by Gay Lussac and Thenard in others from 1808 to 1811. According to Davy, when potassium and phosphorus are heated together, they combine in one uniform ratio of 8 to 3 nearly; and the compound, when acted upon by muriatic acid, given out from .8 to 1 cubic inch of phosphuretted hydrogen gas, resulting from one grain

of the former and $\frac{3}{4}$ of a grain of the latter
substances combined. Also he observed that
half a grain of potassium decomposed nearly
3 cubic inches of phosphuretted hydrogen,
and set free more than 4 cubic inches of hydrogen; the phosphuret seemed to be of the
same kind as the former, or that by direct
combination of the two elements.

Gay Lussac and Thenard combined the
elements by heat; the potassium is scarcely
fused till the phosphuret is formed. The excess of phosphorus sublimes, and the phosphuret is always of a chocolate colour; the
proportions were not ascertained. By treating this phosphuret with warm water, a
quantity of phosphuretted hydrogen was uniformly given out, about 40 per cent. more
than the hydrogen which would have been
yielded by the potassium alone in water.
But if the phosphuret was treated with dilute
acid instead of water, then less gas was given
out than otherwise; and the stronger the acid
the less gas, so as sometimes to reduce the
gas in volume to that yielded by potassium
alone. They also found, as Davy had done,
that potassium heated in phosphuretted hydrogen decomposed it, uniting with the phosphorus and producing the same compound as
in the direct way.

The results of Davy and the French chemists appear to be discordant; but I apprehend they may be reconciled. It appears
probable from both, that the phosphuret of
potassium must be a compound of one atom
of each, or 35 potassium and $9\frac{1}{3}$ phosphorus;
that is, 100 potassium and 27 phosphorus
nearly. Now in Davy's method of treating
the compound with acid, it is most probable
that the atom of potassium takes one of oxygen to form potash, and the atom of phosphorus takes one of hydrogen to form one of
phosphuretted hydrogen; but 3 volumes of
pure phosphuretted hydrogen contain 4 volumes of hydrogen, (see page 178); and Davy
obtained nearly $\frac{3}{4}$ of the volume of gas which
the potassium alone would have produced,
which therefore accounts for the fact as stated by him.

On the other hand, the French chemists by
treating the phosphuret with hot water, probably determined the resolution in this way:
the potassium resolved the water into oxygen
and hydrogen, the last of which was liberated in a free state, and of course produced
the usual volume; the phosphorus also resolved the water into oxygen and hydrogen, one
half of it taking the oxygen to form phosphorous acid, and the other half taking the

hydrogen to form phosphuretted hydrogen, which of course would produce phosphuretted hydrogen amounting to $\frac{3}{8}$ of the volume of free hydrogen or 38 per cent. nearly, which would make up the volume of gas to 138, or nearly 140, as observed by them. It is not unlikely that 2 or 8 per cent. of hydrogen might be added by the further decomposition of water by the phosphorous acid, in order to make it into phosphoric acid.

19. *Phosphuret of sodium.*

No particular experiments having been detailed of this compound, we must infer it is similar to the last mentioned, and consists of one atom of sodium, 21, and one atom of phosphorus, $9\frac{1}{3}$; that is, 100 sodium and 44 phosphorus nearly.

20. *Phosphuret of bismuth.*

If we may judge from M. Pelletier's experiments, bismuth has but a weak affinity for phosphorus. By projecting portions of phosphorus amongst melted bismuth, he succeeded in uniting some of it to the metal; he es-

timates the quantity at 4 per cent.; whereas
by theory it ought to be 15 per cent. supposing them to unite atom to atom.

I do not find that the salts or oxide of bismuth are materially affected by phosphuretted
hydrogen water.

21. *Phosphuret of antimony.*

Phosphorus may be combined with antimony, according to Pelletier, by the same
means as with the other metals. The phosphuret has a white, metallic appearance and
lamellar fracture. The ratio of the elements
was not determined. By theory supposing
one atom to unite with one, it would be 40
to $9\frac{1}{7}$, or 100 antimony to 23 phosphorus
nearly.

Phosphuretted hydrogen water seems to
have no effect on the salts or oxide of antimony.

22. *Phosphuret of arsenic.*

From the experiments of Margraff and
Pelletier, it seems probable that phosphorus
unites both with arsenic and its oxide. By
distilling a mixture of equal parts of arsenic
and phosphorus in a carefully regulated heat,
Pelletier obtained a residuum of a black shin-

ing substance, containing a good proportion of phosphorus. The same was obtained in the humid way, by keeping phosphorus in fusion on arsenic under water for some time. The phosphuretted oxide may be obtained by distilling phosphorus and the white oxide of arsenic together, the phosphuretted oxide sublimes mixed with arsenic and phosphorus in a separate state. It is of a red colour. The proportions in neither case were ascertained. It is probable that the compounds are of the most simple kind, or one atom to one; in that case we shall have 21 arsenic and $9\frac{1}{3}$ phosphorus, or 100 arsenic and 44 phosphorus for phosphuret of arsenic; and 28 oxide and $9\frac{1}{3}$ phosphorus, or 100 oxide and 33 phosphorus for phosphuretted oxide.

No precipitation is occasioned by phosphuretted hydrogen water in solutions of arsenic.

23. *Phosphuret of cobalt.*

Cobalt unites with phosphorus in the direct way as well as by being heated with phosphoric glass. The colour of the compound is a blueish white; it is brittle and crystalline in the fracture. The metal acquires 7 per cent.,

this is below the theoretic quantity, which is
25 per cent. if the atom of cobalt be 37.

Solutions of cobalt give no precipitate by
phosphuretted hydrogen water.

24. *Phosphuret of manganese.*

This compound may be formed like the
preceding ones. It is of a white colour,
brittle and of a granular texture. It is not
liable to be altered by the air like the pure
metal. The proportions of the compound
Pelletier did not determine. Reasoning from
theory, it should consist of 25 metal and $9\frac{1}{3}$
phosphorus; or 100 metal and 37 phos-
phorus.

The salts and oxide of manganese are not
sensibly affected by phosphuretted hydrogen
water.

The combinations of the remaining metals
with phosphorus can scarcely be said to have
been investigated.

SECTION 16.

CARBURETS.

On the supposition that metals combine with charcoal, the appropriate names for the compounds would be *carburets* of the respective metals. This combination, if it exist at all, seems very rare, that with iron being the only one generally acknowledged. No combinations of carbone with the earths and alkalies, have, as far as I know, been noticed; and those with the elements ox gen, hydrogen, sulphur and phosphorus have been described in the former volume. Since that was printed an ingenious experimental essay on the " *Sulphuret of carbon* or alcohol of sulphur," has been published by Berzelius and Dr. Marcet. Some account of this compound, under the name *carburetted sulphur*, has been given (vol. i. page 462); but the additional information is of sufficient importance to require notice here. The pure liquid is of sp. gr. 1.272 ; and the elasticity of its vapour at 66° is equal to 10.76 inches. It burns with a blue flame and sulphureous odour, without sensibly depositing water on cold glass exposed to the fumes. It has an acrid, pungent taste, and a nauseous smell, differing from sulphuretted hydrogen. By various experiments it was found to consist of sulphur and carbone in the ratio of 85 to 15 nearly; that is, 2 atoms of sulphur to 1 of carbone. From other experiments it did not appear to contain any hydrogen.

From some experiments I made in June 1818, on the combustion of the vapour of carburet of sulphur in oxygen gas, I was led to suspect at least, that an atom of hydrogen attaches to the two of sulphur and one of carbone in its constitution. But not having an opportunity to pursue the subject, I merely make the observation for future experience to determine upon the question.

1. *Carburets of iron.*

There are three distinct substances which are now commonly believed to be constituted of carbone and iron, known by the names of *Plumbago,* or black lead, *Cast Iron* and *Steel.*

Plumbago is a natural production, found in greatest perfection in the Borrowdale mine, near Keswick, Cumberland. It is chiefly used in making pencils.—It seems to be constituted of carbone and iron by the concurrent experience of all who have examined it : but the proportions are not uniform, some having found 10 and others only 5 per cent. of iron in it. From this circumstance it would seem doubtful whether iron is an essential element. ?As carbone is known to be exhibited in various forms of aggregation, it is not improbable that plumbago may be one of those forms; it is evidently not a mere mixture of common charcoal and iron, or its oxide.

Cast iron or crude iron is the metal when first extracted from the ore; it usually contains carbone, oxygen, phosphorus and silica, in small proportions, with perhaps other earths occasionally. It cannot be considered as having these elements united in definite proportions; for they vary much, and probably give to crude iron its several modifications.   Cast iron contains about 80 per cent. of its weight of iron in a state capable of solution in dilute sulphuric acid, and yielding a corresponding quantity of hydrogen gas. The residue, in a specimen I examined, was nearly as magnetic as iron itself. When treated with boiling muriatic acid, the insoluble part was reduced to $2\frac{1}{2}$ per cent. upon the original weight of the iron, and some hydrogen gas given out. It was then about as magnetic as the common black oxide of iron; when heated it assumed a glowing red and lost nearly $\frac{1}{2}$ grain; it was still magnetic, and boiling muriatic acid extracted more iron from it.

The hydrogen gas from dilute sulphuric acid and cast iron contains no carbonic acid in my experience; neither does it yield any when exploded with pure oxygen gas,

The small residuum after treating cast iron with acids was found by Bergman and others to resemble plumbago, being constituted chiefly of carbone and iron.

From the above it would seem that cast iron consists chiefly of pure iron, with the addition of very small proportions of oxygen and carbone; the oxygen may be about 1 per cent. and the carbone about 2. These proportions, though sufficient to modify the properties of iron to a certain extent, can scarcely be considered as constituting cast iron a homogeneous mass,

Steel. This most important modification of iron has engaged the attention of many chemists and metallurgists. It may be procured, but not equally pure, by different methods. One is to keep the cast iron for a considerable time in fusion and in a very high degree of heat; whilst its surface is covered with melted scoriæ, so as to preclude the contact of the atmosphere with the iron. This, it is conceived, gives time for the carbone and oxygen to combine and escape in the form of carbonic acid. This steel is of inferior purity.

Steel of cementation is made by stratifying bars of pure iron with charcoal powder in large earthen crucibles, carefully closed up with clay. These are exposed to a high degree of heat in a furnace for 8 or 10 days. This is called *blistered* steel, from the appearance of blisters on its surface.

Cast steel is made from blistered steel by breaking the bars and putting them into a large crucible with pounded glass and charcoal. The crucible is closed with a lid of the same ware and placed in an air furnace. When the fusion is complete the metal is cast into ingots. This is the most valuable and probably the purest steel.

When steel is heated red and plunged into cold water, it is *hardened;* that is, it becomes much harder than iron or than steel without this operation. Hardened steel is brittle, and cannot be extended by the hammer or corroded by a file till it is again softened by being heated and then gradually cooled.

One of the most remarkable properties of hardened steel is that of being *tempered,* as it is called ; by which it is adapted to the different purposes of the manufacturing artists. Tempering consists in heating the hardened steel till it acquires a straw colour for edge tools, a blue colour for watch springs, and elastic articles in general ; &c. &c.

Hardened steel is qualified to acquire magnetism, and to retain it so as to become a permanent magnet. This power of retaining magnetism distinguishes steel from pure iron.

From the above account of steel, it is evident there is an essential difference between it and pure iron. That difference consists, according to the common opinion, in steel being a *carburet* of iron, or carbone and iron united. The fact of the union of carbone and iron in the formation of steel does not seem to me satisfactorily proved. Mr. Collier asserts that iron gains about $\frac{1}{180}$th of its weight by being converted into steel.* But Mr. Mushet found that though steel gains weight upon the iron when copiously imbed-

* Manchester Memoirs, Vol. v. page 120.

ded in charcoal, yet it loses weight if the charcoal is only $\frac{1}{90}$ or $\frac{1}{100}$ of the weight of the iron.* The same ingenious gentleman seems to estimate the carbone in cast steel, from synthetic experiments, to be $\frac{1}{100}$th of its weight.

From analytic experiments, however, there does not appear reason to believe that steel contains so much, if any charcoal Pure steel dissolved in dilute sulphuric acid gives hydrogen gas containing no carbonic acid nor oxide, neither is there any appreciable residuum of any kind in general.

On considering all the circumstances, I am inclined to believe, that the properties which distinguish steel from iron are rather owing to a peculiar crystallization or arrangement of the ultimate particles of iron, than to their combination with carbone or any other substance. In all cases where steel is formed, the mass is brought into a liquid form, or nearly approaching to it, a circumstance which allows the particles to be subject to the law of crystallization. We see that great change is made in steel by the mere *tempering* of it, which cannot be ascribed to the loss or gain of any substance,

* Philos. Mag. Vol. xiii.

but to some modification of the internal arrangement of its particles. Why then may not its differences from iron be ascribed to the same cause? It is allowed that steel, by being repeatedly heated and hammered, becomes iron: that is, it should seem, the change of figure disturbs the regular arrangement of the particles. And it may be further observed, in corroboration of the opinion that cast iron is capable of being made permanently magnetic, from its having been in fusion more probably than from its near approximation to steel in its component parts. The most powerful artificial magnets, after being forged of steel, are said to undergo the operation of steelifying again, before they are hardened finally to receive the magnetic virtue.

SECTION 17.

METALLIC ALLOYS.

When two or more metals of different specific gravities are melted together and intimately mixed, they frequently enter into chemical union and form a new compound, called an alloy of the metals. These alloys

often possess important properties which
their constituents singly do not, and hence
become valuable acquisitions to the arts.
The metals thus combined may be fused together in any proportion ; but if one of them
greatly exceed the other in specific gravity,
their intimate union is sometimes rendered
difficult and even impracticable, partly from
the weak affinity and partly from the gravi-
tating principle causing the metal of least
specific gravity to arise to the surface.

Notwithstanding this union of metals in
seemingly indefinite proportions, there are
only a few proportions in which the alloys
possess peculiar excellences so as to entitle
them to the attention of artists. These proportions have in many instances been discovered by experience ; and it only remains
for theory to point out the reason for such
proportions, and to suggest other proportions which may bid fair to possess desirable
qualities, and thereby diminish the unsuccessful attempts for improvement in these
combinations.

That the metals thus alloyed constitute
true chemical compounds and not merely
mechanical mixtures, may be inferred from
the change made in their primary qualities :
such as

1. *Tenacity, hardness,* &c. Some alloys are much superior to their ingredients in tenacity and hardness, whilst others affect a kind of medium between them. This last is often the case too in regard to ductility and malleability.

2. *Fusibility.* Several alloys fuse at temperatures intermediate between the fusing temperatures of their ingredients, but mostly lower than the mean ; there are others which fuse below the temperature of the lowest, and few if any require a temperature above the mean for their fusion.

3. *Colour.* In many cases the colour of alloys is such as would be produced by the mixture of the colours of the metals; but in others, remarkably different ; for instance, the alloys of copper and zinc,—forming the various kinds of brass.

4. *Specific gravity.* This is not always what might be inferred from a mixture of the two ingredients. Sometimes it is greater and other times less ; but this is not a decisive mark of chemical union, as the same metal varies in specific gravity, by hammering, rolling, tempering, &c. very considerably. Besides, it is more than probable that the differences said to have been observed, have in some instances arisen from inaccurate expe-

riments ; as it is a delicate operation to find
the specific gravity of small pieces of metal
with sufficient precision for comparisons of
this kind.

Many of the simple metals, when fused
and exposed to the air for some time, without a covering of charcoal, or some similar
principle, acquire less or more of oxygen,
and retain it even in a fluid state, as is proved
from Mr. Lucas's interesting communication
in the 3d Vol. of the Manchester Society's
Memoirs (new series). Hence by frequent
fusions of the same metal its quality becomes
impaired in regard to tenuity and other properties.

This is more eminently the case with regard to alloys. Thus, zinc at the temperature in which brass melts is combustible ; and
hence a portion of it escapes by combustion.
Hence the proportions of brass are changed
less or more at each fusion, unless fresh zinc
be added. The same observation applies to
alloys of copper and tin with regard to the
tin. The mixtures of lead, tin, bismuth and
other soft and easily fused metals, are still
more remarkable in this respect. They
should be fused under a cover of oil or tallow in order to keep them of the same proportions ; otherwise, some of them, parti-

cularly the tin, is liable to great oxidation, and no two successive fusions will present the same alloy. Hence in some degree the use of fluxes in metallurgy which serve to cover the surface of the metals and prevent oxidation from the atmosphere.

When an alloy is made, it seldom happens that the metal is perfect and compact the first fusion; it is more or less porous, especially when the two metals fuse at very different temperatures. By a second fusion, which usually takes place at a much lower temperature than that requisite for the first, the metal becomes compact and free from pores. This is particularly the case with speculum metal; and I have little doubt it is so with regard to many other alloys.

Alloys of Gold with other Metals.

Gold unites with many of the metals by heat, and forms various alloys, on which it may be proper to make a few remarks.

1. *Gold with platina.* Platina in a small proportion changes the colour of gold towards white. 1 part to 20 gold makes it much paler. 1 to 11 gives it the colour of

tarnished silver.   1 part with 4 of gold has
much the appearance of platina. The colour of gold does not predominate till it becomes $\frac{8}{9}$ of the alloy. The alloy of 1 platina
and 11 gold is very ductile, and elastic when
hammered. *Lewis. Klaproth. Vauquelin.*

2. *Gold with silver.* These two metals
may be combined in almost any proportion
by fusion and proper treatment. Homberg
found that when equal parts of gold and
silver are kept in fusion for a quarter of an
hour and then cooled, there were two masses,
the uppermost pure silver, the undermost an
alloy of 5 parts gold and 1 silver. I part
silver to 20 gold produces a sensible whiteness in the alloy. 2 parts gold and 1 of
silver are stated to form the alloy of greatest
hardness; this will consist of 3 atoms of gold
to 1 of silver.

3. *Gold with mercury.*   See amalgams.

4. *Gold with copper.* Gold and copper
form an alloy by fusion together. 11 parts
gold and 1 copper form the alloy used for
gold coin. The copper heightens the colour
of the gold, and makes it harder and less
liable to wear. The current gold coin, however, usually contains both silver and copper,
but the weight of both does not much exceed one twelfth of the whole. According

to Muschenbroeck the maximum of hardness is when 7 parts of gold and 1 part of copper are united. This corresponds nearly to 6 atoms of gold and 1 of copper, the atom of gold being estimated at 66 and that of copper at 56.

Other alloys of gold besides the above standard is that for watch cases, which must contain at least ¾ pure gold. Watch chains, and trinkets, are usually made of inferior alloy, called jewellers gold, which is under no controul. It rarely contains less than 30 per cent. of pure gold.

5. *Gold with iron.* Gold and iron may be united by fusion in various proportions. 11 parts gold and 1 iron form a ductile alloy which may be rolled and stamped into coin. Its specific gravity is 16.885. The colour is a pale yellowish gray approaching to white. The alloy is harder than gold. When the iron is three or four times the weight of gold, the alloy has the colour of silver. This last compound is constituted of 1 atom of gold and 8 of iron nearly. *Lewis. Hatchett.*

6. *Gold with nickel.* Mr. Hatchett fused 11 parts gold and 1 nickel together, and obtained a brittle alloy of the colour of fine brass.

7. *Gold with tin.* Gold combines with

tin and forms a brittle alloy. 10 parts gold
and 1 tin form a pale alloy and less ductile
than gold. One fiftieth of tin does not materially injure the ductility. Heat, up to a
visible red, does not impair the alloy; but
beyond that the tin fuses and the alloy falls
to pieces. *Hatchett.*

8. *Gold with lead.* The effect of uniting
even a very small proportion of lead to gold
is remarkable. When the alloy contains
$\frac{1}{1000}$ part of lead, it is brittle like glass.
The vapour of fused lead in close vessels is
sufficient to injure gold. *ibid.*

9. *Gold and zinc.* These two metals combine in almost any proportion. When 11
parts gold and 1 zinc are alloyed, the compound is of a pale greenish yellow like brass,
and very brittle Equal parts of these metals form a very hard, white alloy, suscepti-
ble of a fine polish. *ibid.* & *Hellot.*

10. *Gold and bismuth.* Gold unites with
bismuth, but the colour is injured and the
ductility of the alloy destroyed by a very
small portion of the latter metal, the same
as with lead. *ibid.*

11. *Gold and antimony.* These metals
combine and produce a brittle alloy, much
of the same kind as those with bismuth and
lead. *ibid.*

12. *Gold and arsenic.* There seems a considerable affinity between gold and arsenic, but the volatility of arsenic in the fusing temperature of gold renders it difficult to bring them into contact. A very small proportion of arsenic makes the alloy brittle, and this property increases with the arsenic. *Hatchett.*

13. *Gold with cobalt.* These unite and form a brittle alloy, even when the cobalt only makes $\frac{1}{60}$ of the compound. *ibid.*

14. *Gold and manganese.* Gold and manganese may be united, and the alloy is very hard and less fusible than gold. One alloy was found to consist of 7 or 8 parts of gold and 1 of manganese. *ibid.*

Alloys of Platina with other Metals.

1. *Platina and silver.* It does not appear very clear that these two metals combine by fusion; at least if they do, the difference in their specific gravities is sufficient to overcome their affinity.

2. *Platina and mercury.* See amalgams.

3 *Platina and copper.* These two metals unite with difficulty by a strong heat and form a malleable alloy. This alloy has been preferred for specula for telescopes, as it is

hard, polishes well, and is not liable to tarnish. *Lewis.*

4. *Platina and iron.* Platina and soft or pure iron do not seem to be eas ly combined by heat, by reason of the infusibility of iron. But it combines with cast iron and steel by heat. The alloy is very hard, and in some degree ductile when the iron forms $\frac{3}{4}$ of the alloy. *ibid.*

5. *Platina and tin.* Equal parts of platina and tin unite by fusion, and form a dark coloured brittle alloy. But when the platina falls short $\frac{7}{9}$ of the alloy, the ductility and whiteness proportionally increase. *ibid.*

6. *Platina and lead.* These two metals may be combined in various proportions by heat; but the compounds are not stable, part of the platina falling down, when the alloy is subsequently melted. *ibid.*

7. *Platina and zinc.* Platina may be combined with zinc, by being exposed to the fumes of the metal as reduced from its ore. Three parts of platina become four of alloy. It is hard, brittle, of a blueish white colour, and easily fusible. *ibid.*

8. *Platina and bismuth.* Platina and bismuth combine readily in a high temperature in almost any proportions. The alloys are brittle. *ibid.*

9. *Platina and antimony.* Platina easily combines with antimony by heat. The alloy is brittle. *ibid.*

10. *Platina and arsenic.* When white oxide of arsenic is projected upon strongly heated platina, an imperfect union takes place with a partial fusion of the mass; it is brittle, of a greyish colour and a loose granulated texture. *Lewis.*

Alloys of Silver with other Metals.

1. *Silver with mercury.* See amalgams.

2. *Silver with copper.* Silver and copper are easily alloyed in any proportion by fusion. The compound is harder than silver, and retains its white colour when the copper is half of the alloy or more.—The silver coin is a compound of $12\frac{1}{3}$ silver and 1 copper, which nearly corresponds to 8 atoms of silver and 1 of copper. The hardest alloy is said to be when 5 silver unite to 1 copper; that is, 3 atoms of silver and 1 of copper.

3. *Silver with iron.* The alloys of silver and iron have not been very minutely examined. The two metals are said to unite by fusion, but the iron still retains its magnetism. The alloy is of a white colour, hard

and ductile. When kept in fusion for some time the two metals separate, but not entirely. These circumstances shew the affinity between silver and iron to be weak.

4. *Silver with tin.* Silver and tin form a hard brittle alloy, which is of little if any use. The modifications arising from various proportions have not been particularly investigated.

5. *Silver and lead.* Silver and lead unite in any proportion and form a brittle alloy of a lead colour. The union is not very intimate; for when urged by heat the lead parts from the silver, as in the process of cupellation.

6. *Silver and zinc.* These two unite and form a brittle alloy of a blueish white colour. The proportions have not been particularly noticed.

7. *Silver and bismuth.* Silver and bismuth readily unite by heat. The alloy is brittle and its colour inclines to that of bismuth.

8. *Silver and antimony.* These metals unite by fusion and form a brittle alloy, which does not seem possessed of any remarkable properties.

9. *Silver and arsenic.* These two metals unite according to Bergman, the fused silver taking up $\frac{1}{14}$ of its weight of arsenic; the

alloy corresponds nearly to 3 atoms silver and
1 arsenic. It is brittle and of a yellowish
colour.

Alloys of Mercury and other Metals: Amalgams.

The alloys of Mercury with the various
metals have been commonly denominated
amalgams.

1. *Mercury and gold.* Gold amalgamates
pretty easily with mercury and forms an
alloy much used in gilding metals. For this
purpose six parts of mercury may be heated
nearly to the ebullition of the liquid, and one
part of pure gold in thin plates may be gradually added. In a few minutes the whole
becomes one fluid mass of a yellowish white
colour. It is constituted of 1 atom of gold
and 2 of mercury. By squeezing it through
leather one half of the mercury is separated
nearly pure, and the other remains combined with the gold, and forms a soft white
mass, consisting of 1 part gold and $2\frac{1}{2}$ mercury nearly, which is the alloy of 1 atom to
1, and may be subsequently used for gilding.
A ready way of making this amalgam I find
is to put 3 parts of gold, precipitated by

green sulphate of iron, to 8½ or 9 parts of mercury; by a few minutes trituration the whole becomes a fine crystalline amalgam.—When this amalgam of gold is exposed to a heat just below red, the mercury sublimes and leaves the gold; hence its use in gilding.

2. *Mercury and platina.* These two metals may be combined, but not very easily, as little affinity seems to exist betwixt them. This is manifest from the circumstance that platina wire may be long immersed in mercury without any sensible effect. An union may be produced by immersing thin platina foil into boiling mercury for some time; also by triturating the ammonio-muriate of platina with mercury and exposing it to a due heat. The proportions have not been determined.

3. *Mercury and silver.* Silver and mercury have a considerable affinity and are easily combined by putting lamina of silver into heated mercury and agitating the mixture. When 1 part silver and 2 mercury are mixed as above, a fluid mass is obtained which being heated to the temperature of boiling mercury, a little mercury evaporates and the remainder crystallizes into a soft white mass, which in time grows hard and brittle. A

higher heat than boiling mercury expels the
mercury. Hence this amalgam may be used
for giving a thin coating of silver to the surface of metals, like that of gold. The compound is evidently one atom of silver (90)
with one of mercury (167).

4. *Mercury and copper.* I have made several unsuccessful attempts to combine mercury and copper.

When a plate of copper is kept immersed
in mercury for some time, the mercury adheres to its surface in a small degree and is
not easily rubbed off; the plate is rendered
brittle by it and the fracture has a brilliant
mercurial appearance; but a low red heat
expels the mercury and the copper resumes
its colour and tenacity, with scarcely any
loss of weight, being only about $5\frac{1}{2}$ per cent.
in two or three trials.

Recently precipitated copper in powder,
dried and triturated with mercury, produced
no union. Neither did Dutch-leaf (which is
copper with a very little zinc) unite with
mercury by trituration. Mercury precipitated from deutonitrate by a plate of copper
gave pure running liquid. The plate of
copper appeared as if it had been immersed
in mercury, was brittle with a shining frac-

ture, but recovered its colour and texture by heat, and lost scarcely any weight.

The method recommended by Boyle was tried : 2½ parts of crystallized verdigris, 2 parts of mercury and 1 of common salt, were triturated together till the mercury disappeared, the powder was then digested awhile with vinegar over a fire and frequently stirred. The mass was then put on a filter and dried. It contained a little fluid mercury, but was chiefly composed of acetate of copper and oxide or muriate of mercury. The liquid contained acetate of copper and muriate of soda.

From the above it is manifest that mercury has some chemical action upon copper; but it has not yet been found, I apprehend, that the two metals unite so as to form a proper amalgam.

5. *Mercury and iron.* These two metals have little if any affinity for each other. I do not know that any chemical combination of them has ever been formed.

6. *Mercury and tin.* These two metals readily combine, especially if assisted by heat. I heated 52 parts of tin and 167 of mercury together, that is, 1 atom of each, till they united in a fluid mass. The amalgam crystallized in about 180°. By hard

pressure in the hand nearly 50 parts of fluid mercury were separated from the amalgam when cool, containing in appearance very little tin.   After this an amalgam was formed of 104 parts of tin and 167 mercury (2 atoms tin to 1 mercury); this congealed about 230°, and remained a hard, dry, crystalline substance, agreeing in appearance with that which adheres to mirrors. For the purpose of silvering mirrors however much more mercury is employed than is indicated by the above proportion; but after the glass is slid upon the tinfoil previously covered with mercury, a great pressure is applied, which expels the superfluous mercury nearly in a state of purity.

7. *Mercury and lead.*   To 90 parts of lead I put 167 of mercury (1 atom of each); they united in a moderate heat and crystallized in about 180°. In a few days the mercury partly separated from the amalgam, and 56 parts were squeezed out, the whole was then put together with 90 parts more of lead (now 2 atoms lead to 1 mercury), and fused together; the amalgam crystallized in about 200°, and remained in a solid uniform mass.

8. *Mercury and zinc.*   When 29 parts zinc and 167 mercury (1 atom to 1) are heated together, they combine and form an amalgam

which crystallizes about 200°. A little of the mercury may be squeezed out when cold. By putting 29 parts more of zinc (2 atoms to 1) we obtain an amalgam which fuses considerably above 200°, and when cooled becomes a permanent hard crystalline mass.

9. *Mercury and bismuth.* When 62 parts bismuth are fused with 167 mercury (1 atom to 1), the compound remains fluid at common temperature, but crystallizes partially by standing; about $\frac{1}{3}$ of the weight may be poured off like fluid mercury. If we put 62 bismuth more to the whole (so as to be 2 atoms to 1), the fluid amalgam crystallizes about 150 or 180° : the mass is soft however and by pressure one may squeeze out about 20 per cent. of a fluid amalgam. If we put 62 more bismuth (so as to be 3 atoms to 1), then the compound crystallizes between 200 and 300° into a darkish coloured granular soft mass which continues without any change. Higher than this of bismuth I have not examined.

10. *Mercury and antimony.* Antimony is said to form a feeble union with mercury, which is soon loosened by time. I made several unsuccessful trials to combine these two metals, which it seems unnecessary to detail, as the compound when formed is no ways interesting.

11. *Mercury and arsenic.* On the authority of Lewis an amalgam of mercury and arsenic may be made by keeping them over the fire for some time and constantly agitating the mixture. It is grey-coloured, and composed of 5 parts of mercury and 1 of arsenic.

Most of the other metals are incapable, as far as is known, of combination with mercury, excepting potassium and sodium considered as metals, which combine with mercury; but these alloys are of little interest, and the proportions have not been particularly investigated.

Triple, quadruple, &c. Amalgams.

Besides those amalgams which are formed of mercury and each single metal, there are others formed of mercury and alloys of two or more metals, which in some instances possess properties differing essentially from mere mixtures.

1. *Mercury with bismuth and lead.* When the amalgam formed of 2 atoms bismuth and 1 of mercury is mixed with that formed by 1 atom of lead and 1 of mercury, in such

proportion that the mercury is the same in both, the two powders, though dry and crystalline at first, soon become a permanently fluid amalgam by trituration. The liquid in running along *drags a tail after it*, and is disposed to separate into portions less and more fluid, but the most fluid part is much inferior to pure mercury in this respect. Specific gravity of the amalgam, 11.

2. *Mercury with fusible metal composed of 7 bismuth, 5 lead and 3 tin.* A mixture of 4 parts fusible metal with 5 parts mercury compose the most fusible amalgam with a minimum of mercury that I have found. It is formed of 2 atoms bismuth, 1 lead, 1 tin and 2 mercury. Its specific gravity is 12.

3. *Mercury, zinc and tin.* This amalgam is found the most effectual for the excitation of electric machines. Mr. Cuthbertson recommends 1 part zinc, 1 tin and 2 of mercury for the plate machine amalgam. But for a cylinder the best amalgam I have made contains more than twice the above portion of mercury. I form an alloy of 58 parts zinc and 52 tin, (2 atoms to 1). To this alloy I add 250 mercury, and fuse the mixture; the liquid mass crystallizes about 222° into a white, moderately hard amalgam. This is pulverized in a mortar and mixed up with $\frac{1}{12}$

of its weight of hog's lard. A small portion
then is spread upon a piece of leather and
applied to the machine when in action. It
is probable however that a harder and less
unctuous amalgam may be better adapted to
the plate machine. This amalgam of mine
consists of 4 atoms of zinc, 2 of tin and 3 of
mercury.

I have tried the amalgams of zinc and tin
separately and find that they answer for electric excitation as well as when combined.
They ought to be formed of 2 atoms zinc
and 1 of mercury (58 parts to 167), and of
2 atoms tin and 1 of mercury (104 parts to
167). If we choose to combine them, we
have only to take 2 parts of the zinc
amalgam and 1 of the tin amalgam and tri-
turate them together.

Bismuth amalgam is not good for electric
excitation; lead amalgam is better; but
they are much inferior to those of tin and
zinc.

Alloys of Copper with other Metals.

1. *Copper and iron.* These two metals may
be united with difficulty by heat; but the
compound possesses no useful property.

2. *Copper and nickel.* A white, hard,

brittle alloy is said to be formed by combining
these two metals. The alloy is scarcely
known.

3. *Copper and tin.* The metals of copper and tin, may be fused together and united
in almost any proportion by skilful treatment; but it is found that only a few of the
proportions constitute alloys possessing properties eminently valuable to the arts,

The alloys of copper and tin are commonly
called *bell-metal;* but they receive more particular names according to the purposes for
for which they are destined, as *bronze, spe-
culum-metal, gun-metal,* &c. those of them
which are yellow are frequently confounded in
common language with brass, as *brass-guns,*
&c. Indeed the ancient Greeks and Romans
seem to have been in possession of these two
alloys, under one and the same name. The
χαλκος of the Greeks, being used for cutting-
instruments, must have signified *bell-metal,*
or the alloy of copper and tin as well as brass,
as indeed is proved by the analysis of them.
The *æs* of the Romans seems also to have
included the same compound. Ancient copper coins too are usually found to contain
tin.

Tin united to copper in certain proportions
gives a surprising degree of hardness and ten-

acity to the alloy, much superior in these respects to either of the ingredients. In other proportions it makes the compound highly sonorous, as in *bell-metal* properly so called. Tin also increases the fusibility of the compound in proportion as it abounds, being itself fusible at the low temperature of 440° Fahrenheit.

The principal varieties in the alloys of copper and tin are enumerated below, beginning with those in which the copper is most abundant. The atom of copper is taken at 56 and that of tin at 52 weight, the hardness of these metals is denoted by 7.5 and 6 respectively, by Kirwan.

(*a*). *Gun-metal.* The alloy for brass guns or cannon is made of 100 parts of copper and 11 or 12 of tin. A small portion of iron is found to improve the metal; this is best added in the state of tin-plate, as it more readily fuses and unites with the metal.* This compound is hard and extremely tenacious, exceeding in this respect any other alloy of the two metals. The addition or subtraction of

* See a very excellent essay on the alloy of copper and tin by M. Dussaussoy, in the Annales de Chimie & Physique. 5—113.

1 or 2 parts of tin materially impairs the tenacity of the alloy. It is constituted of 8 atoms of copper and 1 of tin.

(b). *Alloy for edge tools, printers' cylinders, &c.* The best proportion for this compound seems to be 100 parts copper and 15 or 16 tin. When hammered and tempered duly it is fit for making edge tools not inferior to some kinds of steel. It is a compound of greater density than the preceding, though containing more tin; the grain is fine and the metal free from blisters and suited for turning in the lathe. It seems to be the best alloy of the kind for printers' cylinders; but an analysis which I lately made of come turnings from one of these cylinders gave me much less tin than the above proportion. The alloy (b) is constituted of 6 atoms of copper and 1 of tin.

(c). *Alloy for the Chinese gong, cymbals, &c.* An alloy formed of 100 parts copper and 23 tin, appears from Dussaussoy's experiments to form the compound of minimum density. It is used for making cymbals; and nearly accords with the composition of the Chinese gong. It is formed of 4 atoms of copper and 1 of tin. The Chinese gong analysed by Klaproth was composed of 100 copper and

28.2 tin ; that by Dr. Thomson of 100 copper and 23.4 tin.

(d). Common bell-metal used for casting bells. This alloy is commonly made of 3 parts copper and 1 of tin; but to be in due proportion for 3 atoms of copper and 1 of tin, it should be formed of 100 copper and 31 tin. It is hard, of a white colour, less malleable than the preceding alloys, and more sonorous. A specimen I analysed consisted of 100 copper and 36 tin. The exact proportion of 100 copper and 31 tin is not essential to produce a sonorous alloy.

(e). Speculum metal. This compound has been investigated with great care by opticians. According to Mr. Mudge the best proportion is 32 parts copper to 14.5 tin, but Mr. Edwards finds 15 parts tin, 1 brass, 1 silver and 1 arsenic. The slightest variation in the proportions of copper and tin impairs the metal. The alloy is white, hard and close grained ; it takes a beautiful polish. The use of the minute portions of zinc, silver and arsenic is perhaps to correct the colour of the alloy ; though it seems in several alloys that very minute portions of metals apparently foreign to the alloy, improve the density and texture of the metal. It is remarkable with what precision this alloy

accords with the atomic combinations of 2
copper with 1 tin. By calculation 32 copper
would require 14.8 tin. Mr. Mudge finds
32 copper to 14½ tin, and observes that f ¼
a part more of tin be added the metal is too
hard. Mr, Edwards indeed says 32 copper
and 15 tin; but then he adds 1 part brass,
which containing ⅔ of a part of copper, it
reduces his proportion to 32 copper and
14.7 tin, almost exactly that required by the
theory. When 32 copper and 13½ tin are
combined, Mr. Mudge asserts the metal is
too soft.*

(f). *Copper and tin, equal parts.* This
alloy is of blueish white colour, and of no
particular use that I am acquainted with.
It consists of the union of 1 atom of copper
with 1 of tin.

The other alloys of copper with a higher
proportion of tin appear to be uninteresting,
and have not been objects of much attention.

Not having an opportunity of forming these
alloys synthetically, I contented myself with
the analysis of several of them.

* This author obtained the Royal Society's gold medal
for his essay on the composition, &c. of specula for telescopes. Philos. Transact. 1787.

The mode of analysis I adopted with compounds of copper and tin, is simple and easy.
The alloy is treated with nitric acid, which
dissolves the copper, and on being diluted
with water throws down the tin in the state
of deutoxide. This last is collected on a filtre,
dried, and heated to a low red; then $\frac{26}{33}$ of
this is allowed for the tin (the other 7 parts
being oxygen); and the rest of the alloy may
be considered as copper. But if thought
proper the copper may be thrown down by
immersing a plate of lead in the solution,
which succeeds better than a plate of iron in
nitric solutions of copper.

4. *Copper and lead.* Copper unites with
boiling lead and forms a grey brittle alloy
of granular texture. This alloy being heated
above the melting point of lead, causes the
last metal to run off, leaving the copper
nearly pure. The alloy is scarcely of any
use.

5. *Copper and zinc.* Copper and zinc
combined form *brass,* one of the most useful of all alloys. Though this is a general
name for such combinations, yet several of
the proportions form compounds to which
peculiar names are given, some of which will
be noticed below.

It may be proper to remark that copper is

estimated by Mr. Kirwan at $7\frac{1}{2}°$ in hardness, whilst zinc is $6\frac{1}{4}$. The former metal is highly tenacious and malleable; the latter is brittle and malleable only in a small degree. According to Lewis a very small proportion of zinc dilutes the colour of copper and renders it pale; when the copper has imbibed $\frac{1}{12}$ of its weight, the colour inclines to yellow. The yellowness increases with the zinc till the weight of that metal in the alloy equals the copper. Beyond this point if the zinc be increased the alloy becomes paler and paler and at last white, like zinc.

The tenacity of brass is greater than that of either copper or zinc according to Muschenbroek. His experiments give brass nearly twice as strong as copper, and 18 times as strong as zinc. It seems to me most probable that the tenacity of brass increases with the increase of zinc in the alloy to a certain proportion, when it becomes a maximum, and thence diminishes with the further increase of zinc, but experiments are yet wanting, I presume, to ascertain what proportion of the two metals must be taken to form the alloy of greatest tenacity. The same observation may be made as to the maximum hardness; it is not improbable that the two maxima may be found in different kinds of brass.

The point of temperature at which copper fuses is stated to be 27° of Wedgwood's thermometer, whilst that of zinc is much lower, namely, 680° of Fahrenheit. Common brass is stated to melt at 21° of Wedgwood. It is very probable that all kinds of brass melt at temperatures intermediate between those of copper and zinc; and that the more of zinc the lower will be the fusing temperature; but there have not been direct experiments to ascertain the degrees, as far as I know.

In enumerating the different proportions of such alloys as have come under my notice I shall begin with that containing the maximum of copper, and proceed in gradation to that with the maximum of zinc.

(a). *Brass for the manufacture of plated goods.* This alloy is composed, judging from one specimen I analysed, of 12 atoms of copper and 1 of zinc; or of nearly 23 parts of copper by weight and 1 of zinc. The atom of copper is here estimated at 56 and that of zinc at 29, or very nearly $\frac{1}{2}$ that of copper. This alloy had much the same qualities apparently as copper itself, only a little more yellow.

(b). *Dutch gold, gilding metal.* This is the alloy which may be beaten out into thin

leaves, after the manner of gold-leaf. I have
not been able to find any proportions for this
compound in books. It seems to have been
kept as a secret by the manufacturers. By
analysis however I find it composed of 6 atoms
of copper and 1 of zinc, or nearly 12 parts
copper and 1 zinc by weight. This alloy is
probably the most malleable of all the kinds of
brass. A leaf containing 12 square inches
weighs about $\frac{7}{10}$ of a grain. The colour, as
is well known, makes a good approach to
that of gold. It is the composition used for
making articles to be gilt, as buttons, &c.

(c). *Dipping metal for stamped brass goods.*
This is a well known article of Birming-
ham manufacture. It is an alloy both tenacious and malleable, as is manifest from the
perfection of the articles. It possesses a
beautiful gold colour. A specimen was composed, by my analysis, of 4 atoms of copper
to 1 of zinc; or of 8 lbs. of copper and 1
of zinc; or of 4 lbs. copper and 3 of common brass ; but it is varied according to the
colour wanted.

(d). *Soft, fine coloured brass.* According
to M. Sage, a very fine kind of brass may
be made by mixing oxide of copper, calamine, black-flux and charcoal powder toge-
gether, and fusing the mixture in a crucible

till the blue flame disappears. The brass is
found to weigh $\frac{1}{6}$ more than the copper resulting from the weight of oxide. He says when
the copper retains $\frac{1}{5}$ of zinc the colour is not
so fine ; and the excess of zinc will be burned off by heat, but the zinc cannot be reduced
by burning below $\frac{1}{6}$; so that this appears to
be a natural limit. Hence this compound,
being formed of 6 parts copper and 1 of zinc,
must be constituted of 3 atoms of copper and
1 of zinc.

(e). *Soft brass preferred for watch move-
ments.* There is a kind of brass greatly preferred by watch-makers on account of its
working well with steel. I have not met
with a specimen; but Dr. Thomson has analysed one and found it to consist of 2 atoms
of copper and 1 of zinc ;* or 4 parts copper and 1 of zinc by weight nearly.

(f). *Common hard brass.* This constitutes
the great bulk of brass, as manufactured in
the large way. It is made by exposing granulated copper, calamine, that is, a native oxide
of zinc, and powdered charcoal in mixture to
a red heat for some hours, and then increasing
the heat so as to melt the compound of cop-

* An. of Philos. Vol. 12.

per and zinc, the charcoal having carried away the oxygen of the calamine. The metal is then cast into ingots or plates as may be required. This is called brass of cementation as distinguished from the other species, which are usually made from this by fusion with copper or zinc as the case requires.

It is found that 40lbs. of copper with 60lbs. of calamine yield 60 lbs. of brass; hence a great part of the zinc burns away during the process. The brass thus resulting, consisting of 2 parts of copper and 1 of zinc, is of course constituted of 1 atom of each metal united together.

Common brass is malleable, when cold, like the preceding species; but probably does not possess that property in so high a degree. It seems better adapted for turning in the lathe than any other kind of brass. The specific gravity of this brass before it is hammered or rolled is generally about 8.1 or 8.2 by my experience. When rolled it receives a great increase of density, amounting to .5 according to M. Dussaussoy*, so that what is 8.2 when cast will be 8.7 when rol-

---

* An. de Chim. & Physique. 5—233.

led; or it is condensed nearly $\frac{1}{16}$ of its volume by the operation of rolling. The same author finds that brass is hardened very considerably by rolling, but rendered less tenacious; however by being heated and consequently softened after rolling, it becomes stronger than ever, and nearly of an intermediate specific gravity between cast and rolled brass.

(*g*). *Prince's metal, pinchbeck*, &c. This compound, as far as I can learn, is usually formed by combining equal weights of copper and zinc, or by fusing together 3 parts of common brass with 1 of zinc. According to Lewis the yellow colour of brass is a maximum in this proportion. The alloy is brittle, or at least much less malleable than common brass. I find the composition of *spelter* solder, as it is called, or that used for soldering both brass and copper, to be nearly equal parts of copper and zinc. Hence it appears that 1 atom of copper unites to 2 of zinc to form this alloy.

———————

The other alloys of copper and zinc in which the zinc gradually exceeds the copper, become gradually paler in colour and more

brittle. They do not promise to be of much
utility in the arts, and have not therefore
been very particularly investigated by metallurgists.

Besides the binary combinations of copper and zinc and copper and tin, there are
ternary combinations of these metals, namely,
alloys of copper, zinc and tin. For instance,
the metal of which common white buttons
are made. I had occasion to analyse a specimen of this metal and found it to be constituted of 4 parts copper, 1 of zinc and 1 of
tin; or 4 atoms of copper, 2 of zinc and 1
of tin.

———

It will be proper to subjoin the methods
of analysis which I adopted in regard to
brass. Twenty grains, more or less, of the
particular articles were dissolved in nitric
acid, and the metals were precipitated in the
state of sulphurets by hydrosulphuret of
lime. The copper is thrown down in the
state of a black powder, and the zinc in that
of a white powder turning to grey. Great
care was taken to add the precipitating liquor
gradually in order that the copper might be
obtained distinctly from the zinc. The whole

of the copper is thus thrown down before any of the zinc precipitate appears. The precipitates were collected and dried in a temperature not exceeding 150°, and then weighed. In both cases one third of the weight was allowed for sulphur, and the remaining two thirds were estimated to be metal; which is agreeable to the known constitutions of these sulphurets. Another method I sometimes practised, which also answers very well; namely, to throw down the whole or greatest part of the copper by a plate of lead, then to throw down the lead by sulphuric acid; after this the liquor was tested by hydrosulphuret of lime to precipitate the copper remaining, if any; and lastly to throw down the zinc by hydrosulphuret of lime.

6. *Copper and bismuth.* The alloy is brittle and of a pale colour. It is not much known.

7. *Copper with antimony.* Copper and antimony unite by fusion and form a violet coloured, brittle alloy.

8. *Copper and arsenic.* These metals unite by fusion in a close crucible, the surface of the mixture being covered with common salt to prevent the oxidizement of the arsenic. The alloy is white and brittle, and is known

by the names of *white copper,* and *white tombac.*

9. *Copper and manganese.* These may be united by fusion, and form a red coloured. malleable alloy, according to Bergman.

10. *Copper and molybdenum.* These metals may be alloyed in various proportions, but the compounds exhibit nothing peculiarly re- markable.

Alloys of Iron with other Metals.

1. *Iron with tin.* These two metals are alloyed with some difficulty by fusion in a close crucible. The difficulty seems to arise from the very unequal temperatures at which the metals individually fuse. Bergman al- ways found two alloys when the metals were fused together; the one composed of 21 parts tin and 1 of iron, that is, 10 atoms of tin to 1 of iron; and the other of 2 parts iron, and 1 of tin; that is, 4 atoms of iron and 1 of tin. The first was very malleable, harder than tin and not so brilliant; the second but moderately malleable and too hard to yield to the knife.

The formation of common *tin-plate* is a

proof of the affinity of tin and iron. Thin plates of iron, thoroughly cleaned, are dipped into melted tin, when the tin adheres to the surface of the iron, forming with that metal a true chemical union.

2. *Iron and lead,* &c. Iron combines by fusion more or less perfectly with lead, zinc, bismuth, antimony, arsenic, cobalt, manganese, &c. but the proportions have in few instances been ascertained, and the compounds are generally of little importance.

Alloys of Nickel and other Metals.

Nickel and arsenic. As nickel and arsenic are naturally found in combination, though mostly along with small quantities of other bodies, it is to be presumed that an affinity subsists between them; but I do not know that the proportions have been ascertained in which they unite, or the nature of the alloys.

Alloys of Tin with other Metals.

1. *Tin with lead.* Tin and lead unite by fusion in any proportion. This alloy, according to Muschenbroek, is harder and much

more tenacious than either tin or lead, especially when 3 parts tin and 1 lead are its constituents.

I fused various proportions of tin and lead
together, as per the following table, in order
to find some of the more prominent charac-
teristics of the several alloys. The specific
gravity of the tin was 7.2, that of the lead
was 11.23; and the portions taken were such
as to combine, 1, 2, or more atoms of tin with
1 of lead. The several metals were melted
and the compounds formed under a few drops
of tallow, otherwise the oxidation is so rapid
that the proportions are disturbed and the
quantity of pure alloy is not equal to the
weight of the ingredients. Without this
precaution it is no uncommon occurrence in
small experiments to obtain only 3 parts of
fusible alloy from 4 of metal.

Atoms.		Weights	Sp. Gr. by calculation.	Sp. Gr. by experim.	Fusing Point.
Tin.	Lead.	Tin. Lead.			
1	+ 1	.58+1	9.32	9.17	430°
2	+ 1	1.16+1	8.64	8.79	350
3	+ 1	1.73+1	8.25	8.49	340
4	+ 1	2. 3+1	8.00	8.10	345
5	+ 1	2. 9+1	7.93	8.00	350
6	+ 1	3.47+1	7.81	7.90	360

From the above table it appears that when
1 atom of tin is united to 1 of lead there is
an expansion of volume; but when more than

1 of tin are combined to 1 of lead there is a contraction of volume, or the density is above that by calculation. This increase of density is greatest when 3 atoms of tin are combined with 1 of lead ; and it is not improbable the tenacity may then be a maximum ; though Muschenbroek finds it more tenacious when 3 parts tin are united to 1 of lead, which answers more nearly to 4 atoms tin and 1 of lead ; this opinion is countenanced by the fact that tin is much the most tenacious of the two metals taken singly.

It is remarkable that the fusing point of these alloys is below those of either tin or lead. The lowest of all (340°) is when 3 atoms of tin are alloyed with 1 of lead.

Common pewter, I find, is an alloy of 4 atoms of tin and 1 of lead nearly, and fuses about 345 or 350° This is perhaps the best proportion ; it is hard, tenacious and of a good colour. More of lead would impair the colour, and more of tin would impair the tenacity and increase the expence, though it might improve the colour.

Certain articles for family use, such as teapots, spoons, &c. are made of white metal, which commonly, though I apprehend improperly, goes by the name of *tutenag*. This metal in colour approaches more to silver than pewter does. A spoon of this description I found to be pure tin.

2. *Tin and zinc.* This alloy is easily made by fusion. The metals seem to unite in any proportion. I melted together 29 parts zinc and 52 tin (1 atom of each), and obtained a white hard alloy of about 6.8 specific gravity. When 2 atoms tin and 1 zinc are united the specific gravity is 6.77, which is below the mean. The alloy appears to be very hard and tenacious; and probably might be put to some use.

3. *Tin and bismuth.* These metals readily combine by fusion in any proportion. When 52 parts tin and 62 bismuth are fused together (1 atom to 1), a fine, smooth, hard but brittle alloy is obtained of the specific gravity 8.42. It fuses at 260° Two atoms tin and 1 bismuth give an alloy of 8 specific gravity, which fuses about 320° The alloy of 1 atom tin and 2 of bismuth is of 8.67 specific gravity, and fuses about 260° The alloy of 3 atoms tin and 1 bismuth is of 7.73 specific gravity, and fuses at 350° The alloy of 1 atom tin and 3 bismuth is of specific gravity 9.14, and fuses at 330°

4. *Tin with antimony.* This compound is said to be white and brittle when formed of

equal parts. I did not succeed in uniting the two metals by fusion on a small scale.

5. *Tin with arsenic.* When 15 parts of tin and 1 of arsenic are fused together the alloy crystallizes in large plates like bismuth, according to Bayen. It is brittle and less fusible than tin. This alloy must be composed of 5 atoms of tin and 1 of arsenic, that is, 312 tin and 21 arsenic.

Alloys of Lead with other Metals.

1. *Lead and zinc.* These two metals seem to have a weak affinity. They are easily united, or rather mixed, in any proportion by fusion under a little tallow. But however they may be mixed there is a strong tendency to separate again, which no doubt is occasioned in part by their great difference in specific gravity.

I have fused lead and zinc together in various proportions, from 6 parts lead to 1 of zinc, to 1 part lead to 2 of zinc. The compound usually gives a specific gravity rather greater than the mean; but upon being broken the fracture is often like that of zinc in one part and not so in another; and the ana-

lysis of fragments proves that a great difference exists in their composition. Subsequent
fusion sometimes improves the combination
and at other times the contrary. Six parts
lead and 1 of tin gave a compound as nearly
uniform as any. It was 11 specific gravity,
harder and whiter than lead and had much
the appearance of pewter, that is, the alloy
of tin and lead.

2. *Lead and bismuth.* These metals alloy
well. Three parts lead and 2 of bismuth
unite by fusion and form a tenacious alloy
which fuses about 340°. Muschenbroek
found it ten times stronger than lead. It
grows dark coloured soon by keeping. Its
specific gravity by my observation is 10·85,
which is rather greater than the mean. It is
constituted of 1 atom of each metal, or 62
bismuth to 90 lead.

Three parts lead and 4 bismuth (1 atom
lead to 2 bismuth) fuses at 250°. This is
the lowest temperature at which any alloy
of two metals fuses. With a little tin it
makes the triple alloy which fuses lower than
any other metallic compound, without mercury, as will be shown in the sequel. The
specific gravity of this alloy of lead and bismuth is 10.7, which is greater than the mean.

The alloy of 1 part lead and 2 bismuth

(1 atom of lead and 3 bismuth), fuses at 280°, and is of 0.1 specific gravity, or rather less than the mean.

The alloy of three parts lead and 1 bismuth (2 atoms of lead and 1 of bismuth) fuses at 450 . The specific gravity is 11, or rather greater than the mean.

3. *Lead and antimony.* These two metals combine by fusion in any proportion. The alloy is of a fine grain and is brittle or flexible as the antimony or lead prevails. The principal use of this alloy, I believe, is in the formation of printers' types. The small types require a harder alloy or one with more antimony ; the large types have a greater share of lead as being less expensive. On examination of the different types I find 3 proportions of the alloy principally in use. The smallest types are cast from a mixture which very nearly corresponds with 40 parts of antimony to 90 of lead (or 1 atom to 1). It is hard, has a fracture like steel and is of the specific gravity 9.4 or 9.5 nearly, and fuses about 480 or 500°. The proportions were determined both by analysis and by inference from the specific gravity of the metal.

The middle sized types are made of metal composed of 1 atom of antimony and 2 of

lead, or 40 parts antimony and 180 of lead.
This alloy fuses about 450° or 460° and has
the specific gravity of 10 nearly.

The largest types or letters of 2 or 3 inches
diameter are made of metal composed of 1
atom antimony and 3 of lead, or 40 parts to
270. This alloy also fuses about 450 or 460°,
which is a very remarkable fact. Its specific
gravity is usually 10.22. After several trials
I could not determine whether the fusing
point of this or the preceding alloy was lower;
and equal parts of the two alloys fused together were liquified at the same temperature of
450 or 460°.

All the intermediate sizes of types appear
to be made of one or other of the three preceding proportions or of mixtures of them,
the smaller the type the more of antimony
being required to give the requisite hardness.
The largest types might, I conceive, be made
with a much greater proportion of lead.

When 40 antimony and 360 lead (1 atom
to 4) are fused together, the melting point
is about 470° The specific gravity was
found 10.4, but probably too low from blisters or air bubbles. The alloy was more flexible than the preceding, but brittle with a
fine grained fracture.

Forty parts antimony with 450 lead (1

atom to 5) fused at 490°, and gave 11 specific gravity. This alloy bends and breaks with a fine grained fracture.

Forty parts antimony with 540 lead (1 atom to 6) fused at 510°, and gave 10.8 specific gravity, which in all probability was owing to air bubbles. Now the alloy soft and malleable.

4. *Lead and arsenic.* When lead is fused in contact with the white oxide of arsenic under a film of tallow and stirred frequently, an union of the two metals takes place and the excess of white oxide is partially converted into arsenic and partly driven off, seemingly taking with it a portion of the lead. A considerable portion of the mass assumes the form of a black spongy compound infusible at the temperature. It contains a portion of the lead and is probably a compound of the metals with oxygen. The fusible alloy has the appearance of lead, but is brittle, breaks without bending and exhibits a fracture like that of antimony and lead. The specific gravity of the alloy is 10.6, or more if not saturated with lead. By treating it with an excess of nitric acid it is dissolved, and the lead may be thrown down by sulphuric acid, and the arsenic acid or oxide by lime. In this way I find the alloy is com-

posed of about 9 parts of lead with 2 of arsenic, or 1 atom of each of the metals. The spongy mass treated with nitric acid yields a similar solution, accompanied with a precipitation of oxide of arsenic.

5. *Lead and cobalt.* The alloy of these two metals is not easily obtained, probably from the great difference of the temperature at which they fuse. Gmelin fused 1 part cobalt with 1, 2, 4, 6 and 8 parts of lead respectively. Alloys were obtained of the specific gravities 8.12, 12.28 (query 8.28 ?), —, 9.65 and 9.78 respectively. From these specific gravities it is plain the lead had been in great part dissipated by the heat. For the last or greatest specific gravity corresponds nearly to 2 parts lead and 1 of cobalt.   (An. de Chimie, 19—357.)

Triple Alloys, Solders ; Fusible Metal. &c.

Though it may seem premature to treat of triple compounds in the present chapter, which professedly is limited to compounds of two elements, yet as the triple alloys are few and so immediately connected with the preceding, it will scarcely require an apology for introducing them here.

Soft solders. Solders for plumbers and tin-workers, are required to melt easily, and yet not too low, as they should withstand a heat greater than boiling water. The fusing point of the soft solders is usually between 3 400°. Plumbers' solder I believe is commonly formed by mixing equal parts of tin and lead. I procured a specimen of 8.9 specific gravity, and its fusing point was 380° Probably a more perfect compound would be formed by mixing 104 parts tin with 90 lead (2 atom to 1), which would give a specific gravity of 8.8 and the fusing point 350°.

Tin workers' solder is made rather more fusible than that of the plumbers. A specimen I got from the workmen was 8.87 specific gravity and fused at 345°. A mixture of 3 parts tin and 2 of lead would have formed an alloy of the same fusibility, but the specific gravity would have been 8.6 or 8.7 only. Probably a rather less proportion of tin with a little bismuth entered into the composition.

Fusible Metal. Tin, bismuth and lead are metals which melt at comparatively low temperatures; and it has been shewn that the alloys of any two of them usually melt at lower temperatures than the mean, or even than the lower extreme. By analogy it might be inferred that an alloy of tin and lead fused with one of tin and bismuth, would melt below either of the two ingredients. It has been shewn that proportions of bismuth and lead of easiest fusion are 2 atoms bismuth with one of lead ; this alloy melts at 250° An alloy of 2 atoms of bismuth and 1 of tin melts at 260°; and so does that of 1 atom bismuth and 1 tin. These alloys being much more easily fused than any other proportions of these metals, it is from their combinations we are to expect a still further reduction of the fusing point. In fact, a combination of either of the tin and bismuth alloys, with the lead and bismuth alloy, produces almost exactly the same reduction of the fusing temperature.

Thus if 4 atoms of bismuth, 1 of tin and 1 of lead be fused together, the compound melts in boiling water or below 212° It is equally the case if 3 atoms bismuth, 1 of tin and 1 of lead, are fused together.

The double alloy next to those above mentioned in regard to easy fusion is that of 2 atoms tin, and 1 bismuth. It fuses at 320° This alloy, united to the one of 2 atoms bismuth and 1 lead, gives a compound of 3 atoms bismuth 2 tin and 1 lead, which fuses very

L l

nearly at the same temperature as the above triple alloys.

In reference to weights, the above proportions for the most fusible metals will nearly be,

Bismuth 14 parts — Lead 5 —tin 3

————— 10 —— — —— 5 —— 3

————— 5 —— — —— $2\frac{1}{2}$—— 3

Most of the elementary books have given the proportions of 8 bismuth, 5 lead and 3 tin; or 5 bismuth 2 lead and 3 tin, which nearly agree with some of the above, and give an alloy fusing below 212°.

Wishing to investigate this subject more fully, and it being obvious from the preceding facts that there are only two proportions of tin and lead to be united with bismuth, to produce the desired effect, namely, either 1 atom of tin with 1 of lead, or 2 atoms of tin with 1 of lead, I proceeded as follows:

1 atom tin (52)+1 atom lead (90)+1 atom bismuth (62), were fused together; the fusing point was 270°. The alloy was flexible to a certain degree; and the fracture very small grained. To this alloy 31 grains of bismuth were added successively till it was evident the alloy was growing less fusible; the results were as follows:

1 atom tin $+$ 1 lead $+$ 1 bismuth; fuses at 270°
1 —— $+$ 1 —— $+$ 1½ —— —— 235°
1 —— $+$ 1 —— $+$ 2 —— —— 205°
1 —— $+$ 1 —— $+$ 2½ —— —— 200°
1 —— $+$ 1 —— $+$ 3 —— —— 197°
1 —— $+$ 1 —— $+$ 3½ —— —— 200°
1 —— $+$ 1 —— $+$ 4 —— —— 220°
1 —— $+$ 1 —— $+$ 4½ —— —— 205° semi fluid·
1 —— $+$ 1 —— $+$ 5 —— —— 240° semi fluid·

but it retains a little fluidity down to nearly 200°

From this it appears that 3 parts by weight of tin, 5 of lead, and any proportion of bismuth from 7 to 14 will produce an alloy fusing below 212°; but of these the best is 10 or 11 parts.

Again, 2 atoms of tin were combined with 1 of lead and 3 of bismuth, by gradually adding one half of the tin. The several alloys fused without any material difference at or below 200°. A further addition of tin impaired the property as in the above case with bismuth. I did not think it important to mix 2 atoms of tin and 1 of lead with any other proportion than 3 atoms of bismuth.

APPENDIX.

—➤◦◄—

SINCE the publication of the second part of the first volume, (1810) some important essays on the subject of heat have appeared, which have a direct bearing upon some points of the doctrine on that subject inculcated in the said volume. It may be proper to state the results, with such remarks and reflections as have occurred in the consideration of them.

In the Annales de Chimie for January 1813, also in the Annals of Philosophy, vol. 2, we find a Memoir on the specific heat of different gases, by M. M. De la Roche and Berard. This exhibits a most laborious and refined series of experiments on this most difficult subject. Great merit seems to be due to them, both for invention and execution.

It is unnecessary to describe the particulars of the apparatus and the mode of conducting the experiments, as a description may be found as above referred. It is sufficient to observe that the *calorimeter* used was a copper cylinder of 3 inches diameter and 6 in length, filled with water, and having a serpentine tube 5 feet in length, running through the interior and opening at both ends on the outside of the vessels. By means of this tube a regular current of any gas of a given temperature

(212°) might be passed through the vessel so as to part with its excess of temperature to the water. The quantity of water and the capacity of the vessel for heat were previously determined; and the quantity of heated gas passed through the calorimeter was determinable at any time, as well as the temperature of the water, from the judicious arrangements.

It is easy to see that when an apparatus of this kind is at work, the gas will impart heat, more or less according to its capacity, to the water; and that the temperature of the calorimeter will gradually ascend till it arrives at a maximum; that is, till the refrigerating effect of the surrounding atmosphere upon the calorimeter is equal to the heating effect of the current of gas.

The following Table exhibits the results of their experiments.

	Specific Heat	
	Of the same bulk.	Of the same weight.
Air	1.0000	1.0000
Hydrogen	0.9033	12.3401
Carbonic Acid ...	1.2583	0.8280
Oxygen............	0.9765	0.8848
Azote	1.0000	1.0318
Nitrous Oxide ...	1.3503	0.8878
Olefiant Gas	1.5530	1.5763
Carbonic Oxide...	1.0340	1.0805
Aqueous Vapour	1.9600	3.1360†

† The result for this last article must be considered more uncertain than any of the previous ones, the experiment being more complicated.

They found the specific heats of equal volumes of air of the pressures 29.2 and 41.7
inches of mercury to be nearly as 1 : 1.2396,
differing from the ratio of the pressures or
densities, which is 1 : 1.358.

The above table of the specific heat of the
permanent gases (excluding aqueous vapour)
was corroborated by the results of another
series of experiments in which the principle
was varied a little : namely, to find how many
cubic inches of each gas at a given temperature were required to raise the temperature
of the calorimeter a given number of degrees,
and inferring the capacities for heat to be
inversely as the quantities of gas employed.
The differences in the results were from 1 to
10 per cent, which may be considered small,
in experiments of such delicacy.

The ratios of the specific heats of several
gases being found, it was highly expedient to
find the ratio of the specific heat of water,
and that of some one gas, as common air.
This was effected by passing a small current
of hot water through the calorimeter, and
comparing the effect of this current with that
of the larger one of air, the requisite care
being taken to ascertain the quantity of water
passing in a given time and its temperature
at the ingress. The result of this experiment
was that the specific heat of water is to that

of common air as 1 : .25 nearly. By two other experiments, varied from the above, results not much differing were obtained, so that the average of the three gave, water to air, as 1 : .2669.

Reducing the specific heats of the gases to the standard of water as unity, we have the following Table of the specific heats of equal weights of the respective bodies:

Water 1.0000
Air 0.2669
Hydrogen 3.2936
Carbonic Acid......0.2210
Oxygen 0.2361
Azote 0.2754
Nitrous Oxide......0.2369
Olefiant Gas 0.4207
Carbonic Oxide...0.2884
Aqueous Vapour 0.8474

Before we animadvert upon these results, it will be expedient to give an abstract of the not less interesting experiments of Messrs. Dulong and Petit, on heat, as given in the Annales de Chimie and de Physique, vol. 7 and 10.

These gentlemen begin by an investigation of the expansion of air by heat. The absolute expansion of air from freezing of water to boiling had been previonsly determined by

Gay Lussac and myself to be from 8 to 11 nearly : they however extended the enquiry above and below these points of temperature, namely to those of freezing and boiling mercury. From the temperature of freezing mercury or thereabouts, to that of boiling water, they find the expansion of air to keep pace with that of mercury, as indicated by the common thermometer ; but from the boiling point of water to that of mercury, the latter expands somewhat more in a proportion gradually increasing : as by the following Table.

TABLE I.

Temperature by Mercurial Thermometer.		Corresponding volume of a given mass of air.	Temperature by an air Thermometer, corrected for expansion of glass.
Fahrenheit.	Centigrade.		Centigrade.
− 33°	− 36°	0.8650	− 36°
32	0	1.0000	0
212	100	1.3750	100
302	150	1.5576	148.70
392	200	1.7389	197.05
482	250	1.9189	245.05
572	300	2.0976	292 70
680	M. boil 360	2.3125	350.00

The absolute dilatation of mercury claims their attention. They quote nine authorities for the expansion from freezing to boiling water temperatures; the extremes of these nine are, Casbois $\frac{1}{67}$ of original volume, and mine $\frac{1}{50}$ of the same. They determine it to be $\frac{1}{55.5}$. By doubling and tripling the elevation of the temperature, they made observations from which are deduced the results of the following Table. The dilatations are for each degree of the thermometer centigrade, to which I have added the corresponding ones for Fahrenheit's.

TABLE II.

Temperature by an air Thermometer.		Mean absolute dilatations of mercury.		Temperatures indicated by dilatation of mercury, supposed uniform.	
Fahr.	Cent.	Fehr.	Cent.	Fahr.	Cent.
32°	0°	0	0	32°	0°
212	100	$\frac{1}{9900}$	$\frac{1}{5550}$	212	100
392	200	$\frac{1}{9945}$	$\frac{1}{5525}$	400.3	204.61
572	300	$\frac{1}{9540}$	$\frac{1}{5300}$	597.5	314.15

By a series of observations on the apparent dilatation of mercury in glass vessels, compared with the results in the above tables, they deduce the absolute dilatation of glass for each degree of the thermometer, and the temperature that would be indicated by supposing the uniform expansion of a glass rod

---

* That is, the temperature that would be denoted by mercury inclosed in a vessel having no expansion by heat; or else in one that expanded in the same rate as mercury.

M m

adopted as the measure of temperature as under :

TABLE III.

Temperature by an air Thermom.		Mean apparent dilatation of mercury in glass.		Absolute dilatation of glass in volume.		Temperature by a Thermometer made of glass.	
Fahr.	Cent.	Fahr.	Cent.	Fahr.	Cent.	Fah.	Cent.
212°	100°	$\frac{1}{11664}$	$\frac{1}{6480}$	$\frac{1}{69660}$	$\frac{1}{38700}$	212	100
392	200	$\frac{1}{11480}$	$\frac{1}{6378}$	$\frac{1}{63340}$	$\frac{1}{36300}$	415.8	213.2
572	300	$\frac{1}{11372}$	$\frac{1}{6318}$	$\frac{1}{59220}$	$\frac{1}{32000}$	667.2	352.9

The absolute dilatations of iron, copper, and platina were investigated with great address, from 0° to 100° and from 0° to 300° centigrade; and were found as per Table below, for each degree of the centigrade thermometer.

TABLE IV.

Temp.by the air Therm.	Mean dilatation of iron, in volume.	Temp.by iron rod, therm.	Mean dilatation of copper in volume.	Temp.by copper rod thermometer	Mean dilatation of platina in volume.	Temp.by platina rod thermometer
Cent. 100°	$\frac{1}{28200}$	100°	$\frac{1}{19400}$	100°	$\frac{1}{37700}$	100°
300	$\frac{1}{22700}$	372.6	$\frac{1}{17700}$	328.8	$\frac{1}{36300}$	311.6

Connected with this subject was another important enquiry, whether the capacities of bodies for heat remain constant at different temperatures, or whether they diminish or increase as the temperature advances. In other words, does a body that requires a cer-

rain quantity of heat to raise it from 0° to
100° centigrade, require the same quantity
to raise it from 100° to 200 , and from 200
to 300°, &c.; or does it require less or more
as we ascend ?   This enquiry involves that of
the measure of temperature.   They adopt the
uniform expansion of air, or the air thermometer, as the proper measure, and find the
capacity of iron,

From 0° to 100° = .1098
    0   to 200 = .1150
    0   to 300 = .1218
    0   to 350 = .1255

the capacity of an equal weight of water
being 1.

The following Table exhibits the capacities of seven other bodies according to their
results.

TABLE V.

	Mean capacity between 0° and 100°	Mean capacity between 0° and 300°
Mercury......	.0330	.0350
Zinc	.0927	.1015
Antimony ...	.0507	.0549
Silver.........	.0557	.0611
Copper	.0949	.1013
Platina	.0335	.0355
Glass	.1770	.1900

According to this table the capacities of
bodies *increase* with the temperature in a
small degree: and the increase, though it
would still exist, would be less, if the com·
mon mercurial thermometer were the measure
of temperature.

Also supposing that thermometers made of
these bodies and graduated by immersion in
freezing and boiling water into 100° ; if these
were all immersed in a fluid in which an air
thermometer stood at 300°. Then the rela·
tive temperatures of the several thermometers
would be as under, if measured by the absolute quantity of heat acquired, namely,

Iron	322°.2
Silver ...	329. 3
Zinc	328. 5
Antimony	324. 8
Glass......	322. 1
Copper ...	320. 0
Mercury	318. 2
Platina...	317. 9

From these observations they infer that the
law which has been promulgated for the refrigeration of bodies, cannot be strictly true:
namely, that bodies part with heat in proportion as their temperature exceeds that of the
surrounding medium.

Some animadversions on the general laws
relative to the phenomena of heat, announced
in my elements of Chemical Philosophy (page
13) then follow, together with a table drawn

up to show the discordance between the air
thermometer and the mercurial thermometer,
both being graduated in the manner I proposed in the said elements.   On these points
I may have to remark in the sequel.

The first part of the Essay concludes with
some remarks to shew why a preference should
be given to the air thermometer, or more
strictly, the thermometer whether of mercury
or any other body, supposed to be graduated
so as to correspond with an air thermometer
of equal degrees.

The Second Part of the Essay is on

The Laws of Refrigeration.

Adopting the air thermometer as the most
eligible measure of temperature, Messrs.
Dulong and Petit proceed to investigate the
laws of the refrigeration of bodies, under a
great variety of circumstances, in *vacuo* and
in air or gases of different kinds and densities.
The inquiry abounds with experiments and
observations evincing great skill and acute-
ness; but which it will not suit our purpose
to detail.   It may suffice for us to give a general summary of the Laws deduced by them
from their experiments, at the same time re-
commending all those who feel sufficient interest in the subject to peruse the essay at
large, which exhibits a profound philosophical
train of experiments, the results of which

are illustrated by the aid of mathematical generalization.

"*Law* 1. If one could observe the cooling of a body placed in a vacuum, and surrounded by a vessel absolutely destitute of heat, or otherwise deprived of the power of radiating heat, the velocities of cooling would decrease in geometrical progression when the temperatures diminished in arithmetical progression."

"*Law* 2. The temperature of a vessel containing a vacuum being constant, and a body being placed in it to cool, the velocities of cooling for excesses of temperature in arithmetical progression, decrease as the terms of a geometrical progression diminished by a constant number. The ratio of this progression is the same for the cooling of all kinds of bodies, and is equal to 1.0077."

"*Law* 3. The velocity of cooling in a vacuum for the same excess of temperature, increases in geometrical progression, the temperature of the vessel circumscribing the vacuum increasing in an arithmetical progression. The ratio of the progression is the same as above, namely 1.0077 for all kinds of bodies."

"*Law* 4. The velocity of cooling due to the sole contact of a gas is entirely independent of the nature of the surface of the cooling bodies."

" *Law* 5. The velocity of cooling due to the sole contact of a gaseous fluid varies in a geometrical progression, while the excess of temperature itself varies in a geometrical progression. If the ratio of this second progression be 2, that of the first is 2,35, whatever be the nature of the gas and its elastic force.

"This Law may be likewise announced by saying that the quantity of heat carried off by a gas is in all cases proportional to the excess of the temperature of the heated body raised to the power whose index is 1.233."

" *Law* 6. The cooling power of a gaseous fluid diminishes in a geometrical progression, when its tension itself diminishes in a geometrical progression. If the ratio of this second progression is 2, the rate of the first is 1.366 for atmospheric air; 1.301 for hydrogen; 1.431 for carbonic acid; and 1.415 for olefiant gas."

"This law may also be presented as follows: The cooling power of a gas, all other things being alike, is proportional to a certain power of the pressure. The exponent of this power depends on the nature of the gas, and is for air 0.45; for hydrogen 0.315; for carbonic acid 0.517; and for olefiant gas 0.501."

" *Law* 7. The cooling power of a gas varies with its temperature in such a manner that if the gas can dilate so as to preserve the same uniform tension, the cooling power will be as much diminished by the rarefaction of the gas, as it is increased by its augmentation of temperature; so that definitively it depends only on its tension."

————————————

Another ingenious Essay was published by Messrs. Dulong and Petit, in the Annal. de chimie et de physique, vol. 10, namely, " Researches on some important points of the theory of heat."—One object is to ascertain the specific heats of bodies with superior precision. A table of the specific heats of certain metals, found by their method, is given, together with the weights of the atoms of those metals, and the products of the specific heats and weights of the atoms, as under:

	Specific heats, that of water being 1	Weights of the atoms, that of oxygen being; 1	Product of the weight of each atom by the corresponding capacity.
Bismuth.....	0.0288	13.300	0.3830
Lead	0.0293	12.950	0.3794
Gold	0.0298	12.430	0.3704
Platinum ...	0.0314	11.160	0.3740
Tin	0.0514	7.350	0.3779
Silver	0.0557	6.750	0.3759
Zinc	0.0927	4.030	0.3736
Tellurium...	0.0912	4.030	0.3675
Copper......	0.0949	3.957	0.3755
Nickel	0.1035	3.690	0.3819
Iron	0.1100	3.392	0.3731
Cobalt	0.1498	2.460	0.3685
Sulphur	0.1830	2.011	0.3780

The inference intended from this Table is
pretty obvious, namely, that the atoms or
ultimate particles of the above bodies contain
or attach to themselves the same quantity of
heat, or have the same capacity. This principle the authors think will apply to the simple atoms of all bodies, whether solid, liquid,
or elastic; but they hold it does not apply to
compound atoms. It differs therefore essentially from a suggestion of mine, made eighteen
years ago, (see Vol. I. page 70,) that *the
quantity of heat belonging to the ultimate particles of all elastic fluids, must be the same
under the same pressure and temperature.*
They seem to apprehend, from experience,
that a very simple ratio exists between the
capacities of compound atoms and that of the
elementary atoms. They draw another inference from their researches, that the heat
developed at the instant of the combination
of bodies, has no relation to the capacity of
the elements; this loss of heat, they argue,
is often not followed by any diminution in
the capacity of the compounds. They seem
to think that electricity developes heat in the
act of combination; but they do not deny
that a change of capacity may sometimes
ensue, and heat be developed from this cause.

N n

Remarks on the above Essays.

Results nearly agreeing with those of **De la Roche** and **Berard**, on the capacity of certain elastic fluids for heat, were about the same time obtained by **M. M. Clement** and **Desormes.** (See Jonrnal de Physique, Vol. 89—1819.) Such results, impugning some of the most plausible doctrines of heat, could not be admitted but upon very good authority. I remained doubtful, in some degree, till satisfied by my own experience. I procured a calorimeter of the construction of **De la Roche's**, and to simplify the experiment, instead of forcing a given volume of hot air through the calorimeter to impart heat to the water, I drew, by means of an air-pump, a certain volume of atmospheric (or other air) of the common temperature, through the calorimeter filled with hot water, in order to find how much this process would accelerate the cooling. From several experiments of this kind, I am convinced that the capacity of common air for heat is very nearly such as the above ingenious French chemists have determined. That is, it is about $\frac{1}{7}$ part only of what **Dr. Crawford** deduced from his experiments, and nearly the same part of what I inferred from my theoretic view of the specific heats of elastic fluids. (See Vol. I. pages 62 and 74.)

Indeed M. M. De la Roche and Berard appear to have been puzzled with the admission of their own results. The combined heats of oxygen and hydrogen gases give only ,6335 for the specific heat of water; whereas by experiment the heat of water is found to be 1, notwithstanding an immensity of heat is evolved during the combination of these gases.†

" It is necessary therefore," they observe, " to abandon the hypothesis which ascribes the evolution of heat in cases of combination to a diminution of specific heat in the bodies combined, and admit with Black, Lavoisier, and Laplace, and many other philosophers, the existence of caloric in a state of combination in bodies." I am not aware of any writer that denies the existence of caloric in a state of combination of bodies. Dr. Crawford, who would be thought the most likely to err in this respect, maintains, " that ele-

† By recent experiments I find the heat evolved in the union of oxygen and hydrogen, would raise the temperature of the same weight of water 6500°

mentary fire is retained in bodies partly by its attraction to those bodies and partly by the action of the surrounding heat," and that "its union with bodies will resemble that particular species of chemical union wherein the elements are combined by the joint forces of pressure and of attraction." (On animal heat, 2d edition, page 436.) He is perhaps somewhat unfortunate in his instance in the combination of carbonic acid and water; muriatic acid or ammonia and water would have been more in point.

The truth is, these important experiments shew that in elastic fluids the increments of temperature are not proportional to the whole heat, compared with the like increments of temperature and whole heat in those bodies when in the liquid and solid states.

The specific heats of bodies, it is well known, are determined by means of the relative quantities of heat necessary to raise the temperature of those bodies a certain number of degrees. They are expressed by the ratios of those quantities. If the capacities of the same bodies for heat were permanent at all temperatures, then these ratios would also express those of the whole quantities of heat in bodies. In fact, most authors represent

the specific heats as expressing both the ratio
of the total quantities of heat in bodies, and
of the relative quantities to raise their temperature a given number of degrees; but it
is the latter only which they accurately represent, and the former only hypothetically.

In regard to bodies in the solid and liquid
forms, all experience shews that their capacities for heat are nearly if not accurately
constant within the common range of temperature; it seems therefore not unreasonable
to infer that the whole quantity of heat in
each is proportional to their increments.
When, however, a solid body by an increase
of temperature assumes a fluid form, and absorbs heat without any increase of its temperature, its total quantity of heat is thus increased; and it is contended by the writers
on capacity, that the increments of heat afterwards are increased in the same proportion
as the total quantities. This is probable
enough; but it ought to be proved in several
instances by direct experiment before it can
safely be admitted as a general principle;
more especially now since the analogy in the
case of a liquid becoming an elastic fluid is
found to fail in this particular. As an in-

stance of uncertainty, the capacity of ice to
water has been found as 9 to 10 by one person, and as 7.2 to 10 by others; such wide
difference in the results shows there must be
a difficulty in determining the specific heat
of ice, and that it may even be doubted
whether the specific heat of ice or water is
greatest.

From the foregoing detail of experiments
on elastic fluids, it appears evident that such
fluids exhibit matter under a form in which
it has the greatest possible capacity for heat,
when capacity is understood to denote the
total quantity of heat connected with the
fluid; but if the capacity or specific heat is
meant to denote the quantity of heat necessary to raise the body a given number of degrees of temperature, then the elastic fluid
form of matter is that which has the least
capacity for heat of any known form of the
same matter.   When therefore we use the
terms *specific heat* as applied to elastic fluids
we should henceforward carefully distinguish
in what sense they are used; but the terms
may still be indifferently used in the one or
the other sense as applied to liquids and solids,
till some more decisive experiments shew that
a distinction is required.   Probably the ano-

malies that have occurred in investigations of
the zero of cold, or point of total privation
of heat, are in part due to the want of accordance between the ratio of the total quantities of heat in bodies, and the ratio of the
quantities producing equal increments of
temperature.

The greatest possible quantity of heat
which a given weight of elastic fluid can
contain is when the dilatation of the fluid is
extreme. For, condensation, whether arising
from mechanical pressure or from increased
attraction of the atoms of matter for each
other, tends to dissipate the heat, by increasing its elasticity. Hence increase of temperature, at the same time that on one account it increases the absolute quantity of
heat in an elastic fluid, diminishes the
quantity on another account by an increase
of pressure, if the fluid be not suffered to
dilate. This is well known from the fact
that condensation produces increase of temperature in elastic fluids.

When it is considered that all elastic fluids
expand the same quantity by the same increase of temperature, it might be imagined
that all of them would have the same capacity, or require the same quantity of heat to

produce that expansion. The results of De
la Roche and Berard do not seem to admit
of this supposition, though the differences of
the capacities of elastic fluids of equal volumes are not very great. There is a remarkable difference too between their results
and those of Clement and Desormes, in regard to hydrogen gas: namely, .9033 and
.6640; also in carbonic acid gas, 1.2583
and 1.5. The subject deserves further investigation.

In reference to the experiments of Dulong
and Petit, on the relative expansions of air
and mercury by heat, I have no doubt their
results are good approximations to the truth.
My former experiments were chiefly made
in temperatures between 32° and 212°, and I
found, as General Roi had done, the expansion of air to be somewhat greater in the
lower half than in the upper half of that interval, compared with mercury. On a re-
petition of the experiments, I think the difference is less than I concluded it to be, and
I find that the like coincidence of the air
scale and mercurial, continues down to near
freezing mercury; at least the difference will
not be so great as my new table of temperature makes it at page 14. I have made

some experiments on the expansions of air
above 212°, which lead me to adopt the results of Dulong.   On a comparison of the air
and mercurial thermometer upon the laws
which I pointed out, namely, the former ex-
panding in geometrical progression to equal
intervals of temperature, and the latter ex-
panding as the square of the temperature
reckoned from its freezing point, it appears
that in the long range of 600° from freezing
water to boiling mercury, the greatest deviation of the two thermometers does not exceed
22°.   However, the great deviation of the
scales between the temperatures of freezing
water and freezing mercury, is sufficient to
shew, as Dulong and Petit have observed,
that their coincidence is only partial.   Like
the scales of air and mercury, which are so
nearly coincident from −40° to 212° that
scarcely any difference is sensible, though no
one doubts of its existence; yet afterwards
the differences become obvious enough, and
the greater the farther we advance.

Expansion of Mercury.   See page 34, vol.
I.   I have overrated the expansion of glass
bulbs (as will be seen presently,) and hence
that of mercury; my expansion of mercury
corrected on account of the glass, will be
$\frac{1}{57}$ nearly, which leaves it still greater than

Dulong's. The 2nd table of Dulong is valuable, on account of its affording us information of the rate of expansion in the higher
degrees of temperature, from a given or
standard air thermometer.

Iron, Copper, and Platina.

Expansion of Glass.—By the 3rd Table of
Dulong and Petit, it appears these ingenious
chemists found the expansion of glass for
180°, or from 32° to 212°, very nearly the
same as had been determined previously by
Smeaton and others. It also expands in-
creasingly with the temperature, whether it
is estimated by the air or mercurial standard.
This was observed by Deluc, but more ex-
tensively by the present authors. The expansions of iron, copper, and platina, from
32° to 212° as detailed in the 4th table,
agree nearly with the results of others; but
the expansions in the higher part of the scale
manifest some remarkable facts not before
known. Platina not only expands the least
of the above bodies, but its expansion is almost equable; iron expands more than glass
and less than copper, but the most unequally
of any one, the expansion increasing rapidly
as the temperature advances. These facts

explain some others which have fallen under
my observation. I was formerly surprised to
find glass and iron expand so nearly alike
(see vol. I. page 31); but it now appears
that iron increases more slowly in proportion
than glass about the freezing point. More
recently I procured a small thermometrical
vessel of platina to contain water like those
described at page 31, vol. I, and having
filled it and treated it as the other metallic
vessels, I was again surprised to find that the
apparent greatest density of water in this
vessel was at 43°, whereas I expected to have
found it below 42°, the point for glass vessels.
This observation, in conjunction with Dulong's, shews, that platina expands more
than iron at low temperatures, though for a
range of 300° the whole expansion of the
platina is to that of the iron as 2 to 3 nearly.
Hence the error (for I now consider it as
such) which I was led into with respect to
the expansion of glass bulbs, (see vol. I. page
32) and subsequently into that of the expansion of mercury above-mentioned. It is not
the expansion of glass which approaches that
of iron, but it is the reverse, which occasions
the two bodies to meet so nearly in the table,
page 31. This consideration will affect the
point of greatest density of water also; for,

the less the expansion of iron and glass, the
nearer will be the points of real and apparent
greatest density of water, contained in vessels of those materials. My observations on
brown earthenware are scarcely to be relied
upon from the difficulty of making such vessels
water tight; but the common white ware I
have verified repeatedly since the publication
of that table, and am satisfied the point of apparent greatest density, is at or near 40° in
such vessels; hence the real maximum density
of water must be below 40°. I am inclined to
adopt 38° as the most proximate degree.

Capacities of bodies for heat. In the 5th
table of Dulong, we have the specific heats
of glass and of six metals, determined between freezing and boiling water : that of iron
is given before. So far the question does not
involve that of the measure of temperature.
Their results afford no striking differences
from those previously determined ; however,
it is desirable to find a greater accordance
amongst philosophers in this respect. The
experiments which give the specific heats
between 0° and 300° centigrade, are original
and interesting. The results go to shew that
the capacities of bodies increase in a small
degree with the temperature. But supposing
that these results may be relied upon as ac-

curate (which can scarcely be affirmed of any former ones) still the character of them may be changed by adopting a different measure of temperature.

The Essay of M. M. Dulong and Petit, in the 10th vol. of the An. de chimie (see An. of Philosophy, vol. 14th, 1819) manifests great ingenuity. It does not appear, however, so fortunate either in theory or experiment as the former one. It would be difficult to convince any one, either by reasoning or by experience that a number of particles of mercury at the temperature of —40°, whether in the solid, liquid, or elastic state, have all the same capacity for heat. Indeed the experiments of De la Roche and Berard, if they are to be credited, demonstrate the inferior capacity of condensed air to rarefied air; and if the same body changes its capacity in the elastic form, it may well be concluded that all the three forms have not the same capacity. M. M. Dulong and Petit have themselves shewn, in their former essay, (see page 276) that solid bodies vary in their capacities for heat, and that scarcely any two bodies, vary alike; hence it is impossible that the product of the weight of the atom and spefic heat of the body should be a constant quantity. Their specific heat of certain metals differ greatly from what is found by others. For instance, they make the specific heat of lead .0293; the lowest authority I have seen is Crawford, .0352, and the highest Kirwan .050; from repeated trials I have lately found it, upon an average, .032. The weights of some of the atoms in their table, differ materially from what are commonly received; for instance, bismuth is 13.3 instead of 9; also copper, silver, and cobalt, are only half the weights of some authors. The gases too are unfortunate examples. Oxygen gas gives a product of .236 instead of .375; azotic gas gives a product .1967, if oxygen be to azote as 7 to 5, but a product of .393 if oxygen be to azote as 7 to 10: by Dr. Thomson's ratio of oxygen to azote, 4 to 7, the product will be .482, very different from .375. Hydrogen will give a product of .47 or .41 instead of .375. All these differences, it may be said, are occasioned by errors in the specific heats of the gases; but if errors of this magnitude can still subsist after all the care that has been taken, we shall scarcely know what to trust in experimental philosophy.

If M. Dulong would assume all his simple elements in an elastic state and under one uniform pressure, the hypothesis would then make a part of mine (vol. I. page 70), and there is great reason to believe it would be either accurately true or a good approximation; but to suppose that some of the bodies should be in a solid state, having their particles united by various degrees of attraction, others fluid, and others in the elastic state, without any material modifications of their heat arising from these circumstances, appears to me to be in opposition to some of the best established phenomena in the mechanical philosophy.

Their observations on the specific heat of compound elements, on the relation of the heat developed by combination, as compared with the heat of the elements before and after the combination, &c., are not supported by a detail of actual experience. Heat given out by chemical changes they suppose not to have been previously in a state of combination with the elements. As an argument, the heat given out by charcoal kept in a state of ignition, by a current of galvanic fluid, is adduced. It is true this case is most easily explained, by allowing that the galvanic fluid is in such circumstances converted into heat. But the charcoal does not undergo any chemical change, and therefore this is not a case in point.

All modern experience concurs in shewing
that the heat of combustion is primarily de-
pendent on the quantity of oxygen combining.
The heat evolved by the combustion of phosphorus and hydrogen is very nearly, if not
accurately, in proportion to the oxygen spent.
The heat by the combustion of charcoal is
not in a much less ratio: and I find the heat
in burning carbonic oxide, carburetted hydrogen and olefiant gas is the same as in
burning hydrogen gas, *provided the combining
oxygen is the same.*

One difficulty seems to have occurred to
M. M. Dulong and Petit. They all along
conceive that the specific heats of bodies, that
is, the heats producing equal increments
of temperature, must necessarily be propor-
tionate to their whole heat. This is purely
hypothetical, till established by experiment.
The generality of writers on specific heat
had conceived it almost confirmed by experiment. The results of Delaroche and
Berard have shewn that in elastic fluids the
increments of heat are not proportional to the
whole quantities, but on the contrary are less
when a body is elastic than when liquid.
Indeed some writers have argued this should
be the case; because a body nearly saturated

with another has less affinity for it left.*   It
is plain then that oxygen gas or any other
elastic fluid, may have a small specific heat
in the sense above defined, and yet have an
almost unlimited quantity of heat.   I am not
aware of any one established fact that does
not admit of an explanation upon the hypothesis that heat exists in definite quantities in
all bodies, and is incapable of any change,
except perhaps into one of the other equally
imponderable bodies, light or electricity.

* See Dr. Henry's note, Manch. Memoirs, vol. 5,
page 679.

NEW TABLE

OF THE

Forces of Vapours in contact with the generating Liquids at different Temperatures.

Temperatures by the common Thermometer.	Ether vapour, ratio 2 Spec. Gravity .72. Inches of M.	Sulphuret of carbon vapour, ratio 1.978. Inches of M.	Alcohol vapour, ratio 2.7 Spec. Gravity .82. Inches of M.	Acetic acid vapour, ratio 2.57 Inches of M.	Water, ratio 2.602 Inches of M.
7°	3.75		.193		.11
35	7.5		.560	.27	.29
65+	15.	3.134	1.51	.69	.75
97	30	6.20	4.07	1.77	1.95
133	60	12.26	11.00	4.54	5.07
173	120	24.26	29.70	11.7	13.18
220	240	48.	80.2	30.	34.2
272					88.9
340					231

This is an improved and extended table of the force of vapour, similar to that at page 14, vol. I. It shews that the different vapours increase in force in geometrical progression, to certain intervals of temperature, the same to most or all liquids. These intervals of temperature were presumed in the former table, to be in reality *equal* to one another; but the accuracy of this last notion has been questioned.

TABLE,

Shewing the expansion of air, and the elastic force of aqueous and ethereal vapour, at different temperatures.

Temperat.	Vol. of air.	Utmost force of		Weight of 100 cubic inches of aqueous vapour.
		Aqueous vapour.	Ethereal vapour.	
		Inches of Merc.	Inches of Merc.	
−28°	420			Grains.
−20	428			
−10	438			
0	448	.08		
10	458	.12		
20	468	.17		
30	478	.24		
32°	480	26	7.00	.178
33	481	.27	7.18	.184
34	482	.28	7.36	.191
35	483	.29	7.54	.197
36	484	.30	7.73	.203
37	485	.31	7.92	.209
38	486	.32	8.11	.216
39	487	.33	8.30	.222
40	488	.34	8.50	.229
41	489	.35	8.70	.235
42	490	.37	8.90	.245
43	491	.38	9.10	.255
44	492	.40	9.31	.267
45	493	.41	9.52	.275
46	494	.43	9.74	.284
47	495	.44	9.96	.293
48	496	.46	10.18	.303
49	497	.47	10.41	.313
50	498	.49	10.64	.323
51	499	.50	10.87	.329
52	500	.52	11.10	.341
53	501	.54	11.34	.354
54	502	.56	11.59	.366
55	503	.58	11.85	.378

Temperature	Vol. of air.	Utmost force of Aqueous vapour.	Ethereal vapour.	Weight of 100 cubic inches of aqueous vap.
		Inches of M.	Inches of M.	Grains.
56°	504	.59	12.12	.384
57	505	.61	12.39	.396
58	506	.62	12.66	.402
59	507	.64	12.94	.414
60	508	.65	13.22	.420
61	509	.67	13.51	.432
62	510	.69	13 80	.444
63	511	.71	14.10	.456
64	512	.73	14.41	.468
65	513	.75	14.72	.480
66	514	.77	15.04	.492
67	515	.80	15.36	.509
68	516	.82	15.68	.521
69	517	.85	15.90	.539
70	518	.87	16.23	.551
71	519	.90	16.56	.569
72	520	.92	17.00	.580
73	521	.95	17.35	.598
74	522	.97	17.71	.610
75	523	1.00	18.08	.627
76	524	1.03	18.45	.645
77	525	1.06	18.83	.662
78	526	1.09	19.21	.680
79	527	1.12	19.60	.700
80	528	1.16	20.00	.721

Applications of the above Table.

These tables will be found of great use in reducing volumes of air from one temperature or pressure to any other given one: also in determining the specific gravities of dry gases from experiments on those saturated with or containing given quantities of aqueous or other vapours.

As several writers, and some of considerable eminence, have given erroneous or imperfect formulæ on these subjects, more particularly with regard to the effect of aqueous vapour in modifying the weights and volumes of gases, it has been thought proper to subjoin the following precepts and examples for the use of those who are not sufficiently conversant in such calculations.

The 5th column of the above table, or weight of aqueous vapour, is new, and may therefore require explanation. Gay Lussac is considered the best authority in regard to the specific gravity of steam; but it would be well if his results were confirmed or corrected, as they are of importance. According to his experience, the specific gravities of common air and of pure aqueous vapour, *of the same temperature and pressure,* are as 8 to 5, or as 1 to .625. Now I assume that 100 cubic inches of common air, free from moisture, of the temperature 60° and the pressure of 30 inches of mercury, weigh 31 grains nearly. It is an extraordinary fact that philosophers are not agreed upon the absolute weight of a given volume of common air. Most authors now assume the weight of 100 inches = 30.5 grains, whilst according to my experience it is more than 31 grains. If common air be assumed 31 grains, steam would be $19\frac{3}{8}$ grains for 100 cubic inches, at the same temperature and pressure, could it subsist; but as it cannot sustain that pressure at the temperature of 60° we must deduct according to the diminished pressure, the utmost force of steam at 60°being .65 parts of an inch of mercury, we have 30 inches : $19\frac{3}{8}$ grains : : .65 : .420 grains = the weight of 100 cubic inches of aqueous vapour at 60° and pressure .65 parts of an inch; which is the number given above in the table. The like calculation is required for any other pressure : but in addition to this, there is to be an allowance for the temperature from the 2d column: Thus, let the weight of 100 cubic inches of steam at 32° be required. We have 30 inch. : $19\frac{3}{8}$ grs. : : .26 inch. : .1679 grs.; the weight of 100 inches of steam at 60°; then if 480 : 508 : : .1679 : 178 grs. = weight of 100 cubic inches of steam at 32° and pressure .26 parts of an inch, the tabular number required.

Examples.

1. How many cubic inches of air at 60° are equivalent in weight to 100 cubic inches at 45° ?

By the column headed *volume of air* we have this proportion, if 493 : 508 : : 100 inch. : 103.04 inches, the volume required.

2. How many cubic inches of air with the barometer at 30 inches height, are equal in weight to 100 cubic inches when the barometer stands at 28.9 inches ?

Rule. The volume of air being inversely as the pressure, we have, 30 : 28.9 : : 100 inches : $96\frac{1}{3}$ inches the answer.

3. How many cubic inches of dry air are there in 100 inches saturated with aqueous vapour, at the temperature of 50°, and pressure 30 inches of mercury?

Here the formula $\frac{p-f}{p}$ applies, where p denotes the atmospheric pressure at the time, and f denotes the utmost force of vapour in contact with water at the temperature. Hence $p = 30, f = .49$ per table, and we have $\frac{p-f}{p} = \frac{30-.49}{30} = \frac{29.51}{30} = 98\frac{11}{30}$, or

$98\frac{11}{30}$ per cent dry air.

& $1\frac{19}{30}$ vapour.
100.

If the vapour of ether is assumed, then $f = 10.64$, and we have $\frac{p-f}{p} = \frac{30-10.64}{30} = \frac{19.36}{30} = .645$, ...or $64\frac{1}{2}$ per cent dry air.*

$35\frac{1}{2}$ per cent ethereal vapour.
100

4. Suppose we find by trial the weight of 100 cubic inches of common air saturated with vapour at 60°, the barometer standing at 30 inches to be 30.5 grains, and the weight

* The aqueous vapour in this case may be considered as insignificant.

of hydrogen gas in like circumstances to be
2,118 grains; query the weights of 100 cubic
inches of each gas free from vapour, and
their specific gravities, the temperature and
pressure being as above?

If 30.5 : 2.118 : : 1 : .0694 = sp. gr. of
vapourized hydrogen, that of vapourized air
being 1. Subtracting .42 grs. (weight of
vapour per table) from 30.5 grs., leaves 30.08
grains; and subtracting .65 parts of an inch
from 30 inches, leaves 29.35 inches. Hence
100 cubic inches of dry air at the pressure of
29.35 inches, weigh 30.08 grains; and we
have 29.35 : 30 : : 30.08 : 30.746 grains, the
weight of 100 inches of dry air. Again,
subtracting .42 grs. from 2.118, leaves 1.698
grains = weight of 100 cubic inches of hydrogen of 60° and sustaining the pressure of
29.35 inches; whence if 29.35 : 30 : : 1.698
: 1.736 grains, weight of 100 inches of dry
hydrogen; and 30.746 : 1.736 : : 1 : .05645
= sp. gr. of dry hydrogen, that of dry air
being unity. Or the results may be exhibited
as under:

Weight of 100 cubic inches.		Sp. Gravities.	
Vap. air............30.5	grains	1............	14.4
Vap. hydrogen 2.118 ——		.0694	1
Dry air............ 30.746	grains	1............	17.7
Dry hydrogen 1.736 ——		.05645	1˙

FORMULÆ FOR DETERMINING THE PROPORTIONS OF COM-
BUSTIBLE GASES IN MIXTURES.

It frequently happens, especially in the decomposition of vegetable substances by heat, that the product consists of several combustible gases in mixture, and it is desirable to determine the proportions of each of those which collectively constitute the mixture. The following forms will be found useful for this purpose.

1. *Carbonic oxide and hydrogen.*

Let $x =$ the volume of carbonic oxide, $y =$ that of hydrogen, $w =$ that of mixture, and $a =$ that of carbonic acid, produced by exploding the mixed gases with oxygen over mercury.

Then the carbonic oxide, or $x = a$,

and the hydrogen, or $y = w - a$.

2. *Sulphuretted hydrogen and hydrogen.*

Let $x =$ the volume of sulphuretted hydrogen, $y =$ that of hydrogen, $w =$ that of the mixture, and $g =$ the oxygen spent in the combustion of w.

Then because $x + y = w$, and $1\frac{1}{2}x + \frac{1}{2}y = g$;

we have $x = g - \frac{1}{2}w$, and $y = 1\frac{1}{2}n - g$.

P p

3. *Phosphuretted hydrogen and hydrogen; also carburetted hydrogen and hydrogen, and carburetted hydrogen and carbonic oxide.*

The notation being as above, we have $x + y = w$, and $2x + \frac{1}{2}y = g$ (see page 171): and, $x = \frac{2g - w}{3}$, and $y = \frac{4w - 2g}{3}$

4. *Olefiant gas and carburetted hydrogen.*

The notation being as above, we have $x + y = w$, and $3x + 2y = g$; whence $x = g - 2w$ and $y = 3w - g$.

5. *Carburetted hydrogen, carbonic oxide and hydrogen.*

Let $x =$ carburetted hydrogen, $y =$ carbonic oxide, $z =$ the hydrogen, $g =$ the oxygen spent in the combustion of w volumes of mixed gas, and $a =$ the carbonic acid produced.

$$\text{Then } x + y + z = w,$$
$$x + \tfrac{1}{2}y + \tfrac{1}{2}z = g,$$
$$\text{and } 2x + y \qquad = a.$$

whence we have $x = \frac{2g - w}{3}$, $y = \frac{3a - 2g + w}{3}$ and $z = w - a$.

6. *Olefiant gas, carburetted hydrogen and carbonic oxide.*

Let $a =$ the olefiant gas, $y =$ the carburetted hydrogen, and $z =$ equal the carbonic oxide, $g =$ the oxygen entering into combination, and a = the carbonic acid produced; also w = the whole volume as before.

Then we have $x + y + z = w$,
$$3x + 2y + \tfrac{1}{2}z = g,$$
and $2x + y + z = a$.

whence $x = a - w$,
$$y = \frac{4w - 5a + 2g}{3}$$
and $z = \tfrac{2}{3}(w + a - g)$.

7. *Superolefiant gas,** carburetted hydrogen, and carbonic oxide.*

Let x = volume of superolefiant, y = volume of carburetted hydrogen, z = volume of carbonic oxide, g = the oxygen combining, a = carbonic acid produced, and w = volume of mixed gas.

Then $x + y + z = w$,
$$4\tfrac{1}{2}x + 2y + \tfrac{1}{2}z = g,$$
and $3x + y + z = a$.

Whence $x = \frac{a - w}{2}$,
$$y = \frac{3w - 4a + 2g}{3},$$
and $z = \frac{3w - 4g + 5a}{6}$

* A gas found in oil and coal gas. See Manchester Memoirs, vol. 4 (new series), page 73.

8. Superolefiant gas, carburetted hydrogen, carbonic oxide, and hydrogen.

This is the mixture of gases obtained by a red heat from coal and oil, after being freed from carbonic acid, &c., by the usual means.

This mixture requires a very complicated formula, in consequence of the specific gravities of the gases entering into the calculus. The importance of the subject however may be an apology for the labour.

Let $x =$ vol. of superolefiant, S its sp. gr.

$\quad y =$ vol. of carb. hydrogen, f its sp. gr.

$\quad z =$ vol. of carbonic oxide, c its sp. gr.

& $u =$ vol. of hydrogen, $\quad s$ its sp. gr.

$\quad C =$ specific gravity of the mixture, $y =$ oxygen, $a =$ carbonic acid. and $w =$ whole volume of mixture as before.

Then we have $x + y + z + u = w$

$$4\tfrac{1}{2}\,x + 2y + \tfrac{1}{2}z + \tfrac{1}{2}u = y$$

$$3\,x + y + z = a$$

And $S\,x + fy + cz + su = Cw.$

Whence $u =$

$$\frac{(3S + 5c - 8f)a - (4c - 4f)\,g - (3S + 6C - 6f - 3c)\,w.}{8f + c - 3\,S - 6s}$$

The value of the hydrogen being obtained, it may be subtracted from w, and the remainder will be best divided into three portions, by the preceding formula.

HEAT PRODUCED BY THE COMBUSTION OF GASES.

Subsequent experience to that detailed at page 77, Vol. 1. has furnished the following more correct results of the heat produced by the combustion of pure gases.

Hydrogen, combustion of it raises an equal volume of water	5°
Carbonic Oxide	4½
Carburetted Hydrogen, or Pond Gas	18
Olefiant Gas	27
Coal Gas (varies with the gas from 10° to).........	16
Oil Gas (varies also with the gas from 12° to) ...	20

Generally the combustible gases give out heat nearly in proportion to the oxygen they consume. See note at the end of Vol. 4, new series of the Manchester memoirs.

ABSORPTION OF GASES BY WATER, &c.

This curious subject has attracted much less attention than it deserves. Very little has been published relating to it since the time of Dr. Henry's essays and my own, now more than twenty years ago. The only author I remember is M. Saussure of Geneva, who published a similar essay about twelve years afterwards. See Thomson's Annals of Philosophy, Vol. 6. He investigates the

quantities of gases absorbed by various solid
bodies, in a manner which I do not fully
comprehend ; he then treats of the absorption
of gases by liquids, adverting at the same time
to Dr. Henry's experiments and mine. My
enquiries were principally confined to *one*
liquid, water; but I made a few trials with
others, such as weak aqueous solutions of salts,
alcohol, &c., and observing no remarkable
differences, I concluded somewhat too hastily
that "most liquids free from viscidity, such
as acids, alcohol, &c., absorb the same quantity of gases as pure water." Manchester
memoirs, new series, Vol. 1. M. Saussure
however asserts that there are considerable
differences in liquids in this respect. He
finds sulphuretted hydrogen to be more absorbable by water than Dr. Henry and I did;
in this I find he is right. Water takes about
$2\frac{1}{2}$ its bulk of this gas when pure; and it
seldom had been obtained unmixed with hydrogen when Dr. Henry and I made our
experiments upon its absorption. In regard
to carbonic acid, nitrous oxide, and olefiant
gas, M. Saussure nearly agrees with us; but
his results with oxygen gas, carbonic oxide,
carburetted hydrogen, hydrogen and azote,
would prove that water absorbs twice the

quantities of each that we have assigned. I have no doubt he is wrong in the less absorbable gases. In the case of absorption of mixed gases, Saussure has given four examples, in which he finds the results to militate against my theoretic view, as stated at page 201, Vol. 1.; namely, that water takes the same quantity of each in a mixed state as it would do if they were separate, and in other respects in like circumstances. But I have shewn in the Annals of Philos. Vol. 7, 1816, that his results coincide as near as any one can expect with the views which I have all along taken of this subject.

It will be seen, page 173, that another gas has been found to coincide with olefiant gas in absorbability; namely, phosphuretted hydrogen.

FLUORIC ACID.—DEUTOXIDE OF HYDROGEN.

In treating of Fluoric acid, (Vol. 1, page 277) we came to the conclusion that this acid was probably constituted of two atoms of oxygen, and one of hydrogen, and have figured it accordingly (Plate 5, fig. 38). Subsequent experience however has shewn that deutoxide of hydrogen, though it can be formed synthetically, is not the same thing as fluoric acid. We are indebted to M. Thenard for the discovery of this curious compound, the deutoxide of hydrogen or oxygenated water. An ingenious memoir on the subject was published by him in 1818, in which the formation and the properties of this compound are fully detailed. I had no small satisfaction in 1822, when at Paris, in being obligingly favoured by M. Thenard with a view of the process of the formation, and of the more distinguishing properties of this singular liquid.

The nature of fluoric acid is still enveloped in obscurity. My experience led me to adopt the composition of fluate of lime to be 40 acid and 60 lime per cent. I had not then seen Scheele's admirable essay on the subject. From the 5th section of his 2d. essay on fluor mineral, 1771, it may be deduced that fluate of lime is composed of 72.5 lime and 27.5 acid per cent. In 1809 Klaproth, and near the same time, Dr. Thomson found about $67\frac{1}{2}$ lime and $32\frac{1}{2}$ acid per cent. in fluor spar. They both erred, no doubt, as I did, by not repeating the treatment of the mineral with sulphuric acid often enough. Since then most authors, as Davy, Berzelius, Thomson, &c., agree with Scheele nearly, in assigning 27.5 acid, and 72.5 lime, in 100 parts of fluate of lime. My experience in 1820 gave me 1 per cent less of lime; and Dr. Thomson now finds about 1 per cent more of lime than Scheele's analysis gives.

If we estimate the atom of lime at 24, that of fluoric acid must be about 9, according with the above proportion; this is much below 15, the weight of an atom of deutoxide of hydrogen.

Should Sir H. Davy's view of fluate of lime be found correct, its atomic constitution would be one atom of *calcium*, the metallic substance of which lime is the protoxide, and one atom of *fluorine*, the name he has assigned to the other element, which with hydrogen is supposed to constitute the fluoric acid. The atom of fluor spar would then be 1 atom of calcium, 17, united to one atom of fluorine 16.

MURIATIC ACID.—OXYMURIATIC ACID, &c.

From the articles *muriatic acid* and *oxymu-riatic acid* in the former volume, published now 16 years ago, as well as from the appendix to said volume, in which sundry animadversions are found on the fluctuating opinions entertained in regard to these acids,

Q q

the reader will not be surprised to find some
further addition.

Three notions have been submitted to the
public in the last twenty years in regard to
the nature of muriatic acid. First, the gas
detached from common salt by sulphuric acid
has been thought to be the acid in a state of
purity, and constituted of a certain base or
radical united to oxygen; this was the notion
inculcated in the articles alluded to above.
Second,—it is stated as a fact that when oxymuriatic acid and hydrogen in equal volumes
are united by the electric spark, a volume of
muriatic acid gas is the result equal to the
sum of both the other volumes, and that this
gas perfectly agrees with the gas obtained
from common salt by sulphuric acid; this
suggested the idea that muriatic acid gas is
a compound of what has been called *real* or
dry muriatic acid one atom, and water one
atom. And, third, it is argued, that the element we have called oxymuriatic acid gas, is,
for aught that appears, a *simple* body, and
consequently, that muriatic acid gas is the
real acid, and is constituted as above, of one
atom of hydrogen, and one atom of oxymuriatic acid (now called *chlorine.*) It is not
intended here to enter into a discussion of the
arguments and facts adduced in support of

the different conclusions. More experience must be had before all the doubts and difficulties are removed from the subject. But it will be proper to illustrate these different positions by an example. For instance, common salt, muriate of soda or chloride of sodium. By the first notion 50 parts of dry common salt will consist of one atom of muriatic acid gas, 22, and one atom of caustic soda, 28. By the second notion the same salt will be formed of 30 parts of muriatic acid gas, and 28 of caustic soda; but 8 parts of water evaporate when the salt is dried. By the third view common salt consists of oxymuriatic acid, or chlorine and sodium, or the metal of which caustic soda is the protoxide; and 50 parts of salt will consist of 29 chlorine and 21 sodium, or one atom of each.

NITRIC ACID—COMPOUNDS OF AZOTE AND OXYGEN.

Since the account of nitric acid (Vol. 1, page 343) was printed, a change has universally taken place in estimating the weight of the nitric acid atom, and of the proportion of azote and oxygen in the same. This has been effected chiefly by a more correct analysis of nitre than existed at that time. Nitre is now found to consist nearly of 52 parts

acid and 48 parts potash per cent. Hence if
the atom of potash be 42, that of nitric acid
must be 45; for, 48 : 52 : : 42 : 45, nearly.
That is, the nitric acid atom consists of 10
azote + 35 oxygen by weight; or of 2 atoms
of azote (according to my estimate) and 5
of oxygen. There appear to be *two* nitrous
acids; namely, the one which I have des-
ignated by that name, which may now be
called *sub-nitrous*, or as Gay Lussac terms it
pernitrous; and the other what I considered
as *nitric acid* in the former volume, composed
of 1 atom azote, and 2 of oxygen.

Real nitric acid then is that combination
which is effected by uniting oxygen with a
minimum of nitrous gas; or 1 measure of oxygen with 1.3 nitrous gas, (See Vol. 1, page
328). The oxynitric acid, which I was led
to infer from the last mentioned combination,
(1 azote with 3 oxygen) does not appear to
exist. The Table of nitric acid (Vol, 1,
page 355) will require some correction. An
increase of about 4 per cent. should be made,
I apprehend, on the quantities of acid corresponding to the several specific gravities.

Since my former volume of Chemistry was
printed, several essays on the compounds of
azote and oxygen have been published, with
some new and some additional experiments,

the chief of which may be seen in Sir H. Davy's Elements of Chemical Philosophy, the Annales de chimie et de physique, Vol. 1 ; Annals of philosophy, Vol. 9 and 10; and the Manchester Society's Memoirs, Vol. 4, *second series;* also Dr. Thomson's first principles of Chemistry. Notwithstanding all that has been written on the subject, there still appears uncertainty as to the number of combinations formed by these two elements, their relative weights, and the number of atoms in the several compounds.

The results of an experiment I lately made on the decomposition of nitrate of potash by heat seem to be worthy of record, as I am not acquainted with those of any other person who has pursued the experiment to the same extent.—I took an iron retort of 6 cubic inches capacity, and cleaned it as well as I could from carbonaceous matter which it had previously contained, first by heating nitre to redness for an hour or more in it, and then washing it repeatedly with water till the liquid came out tasteless, and only mixed with a little red rust; I then put in 480 grains of purified nitre, and having secured a copper tube to the retort so as to be air tight, the retort was put into a fire and gradually raised to a red heat, and the fire was occasionally

urged with a pair of bellows, in order to
keep up a glowing red on the retort for nearly
two hours; the air was received over water in
jars; the first 4 or 5 inches were thrown
away, and the rest was preserved and transferred to a graduated jar; the products were
examined in successive portions as under,
namely,

Inches.

1 produce,	85 cubic inches,	83 per cent pure	=	70.5
2	5	77	=	3.85
3	25	50	=	12.5
4	6	30	=	1.8

Total 121   Oxygen 88.65
Oxygen 88.65 = 30 grains.

Residue 32.35 = 10 grains.

About 1 per cent on the whole gas was carbonic acid, the rest oxygen and azote, the
weights of which would be nearly as above.

Towards the last the gas came very slowly,
and being of inferior quality, the operation
was discontinued.

The remaining contents of the retort were
diluted with water, and well washed till the
water ceased to shew alkali; the liquid was
then concentrated and gave 1600 water grain
measures of the sp. gr. 1.153. There were
obtained also 64 grains of red oxide of iron
from the washing of the retort, containing
19 grains of oxygen.

The liquid was divided into portions and examined ; the original nitre consisted of 250 grains of nitric acid united to 230 of potash = 480 grains. After the process there appeared to be,

10 grains of carbonic acid united to 21 grains potash.
62 grains of subnitrous acid to   -   84 ....
134 grains nitric acid to   -   125 ....
                                      ———
                                      230

The quantity of carbonic acid was determined by lime-water : the quantity of potash uncombined with nitric acid was found by precipitating it by tartaric acid, and manifested 105 grains of potash in the bitartrate = that combined with the carbonic and subnitrous acids; from which subtracting 21, it was inferred the remainder 84 must have been in union with subnitrous acid, or else with nitrous acid ; the rest of the potash, not being acted upon by tartaric acid, was understood to be combined with nitric acid.

The quantity of subnitrous acid given above, appeared somewhat hypothetical, till it was confirmed by treating a portion of the liquid with oxymuriate of lime solution of known strength ; it was found that 32 grains of oxygen were required to be combined with the subnitrous acid, in order to restore it to the state of nitric acid; that is, when oxy-

muriate of lime, containing that quantity of
oxygen, was added to the liquid, and this
was afterwards rendered acidulous by the addition of sulphuric acid, neither nitrous vapour nor oxymuriatic gas was perceptible;
but a greater or less quantity of the oxymuriate being applied, and the liquid made
acidulous, the fumes of the one or the other
were abundantly manifest.

It remains to account for the oxygen.
There were 250 grains of nitric acid at first
in the nitre; of which 200 grains were oxygen and 50 azote, nearly. One-fifth part of
the oxygen = 40 grains, corresponds to 1
atom of oxygen. Now the whole of the
oxygen derived from the nitre in the course
of the experiment, seems to be 30 grains in
gas, 7 grains in the carbonic acid, and 19
grains in the iron oxide, together equal to
56 grains. Now the azote and oxygen in the
gas collected, were very nearly in the proportion of those elements in nitric acid; so
that a portion of the acid (about $\frac{1}{6}$) might
be considered as completely decomposed,
whilst the rest was only losing a small part
of its oxygen : this is remarkable, and I think
indicates that the carbonic acid (formed from
the carbon of the retort, or from the adhering
carbon) unites to the potash, expelling the *ni-*

trous acid, which is immediately decomposed
into its elements azote and oxygen. This would
not however account for the whole of the
azote: for, 40 grains of nitric acid would be
united to 37 potash; whereas we find only
21 potash with carbonic acid; and I cannot
believe that an error in the estimate of carbonate of potash could exist to that amount.
The fact, however, was, that the elements of
40 grains of nitric acid were found in the
evolved gas, and hence we have to account
for the remainder 210 grains.   From this
there appears to have been expelled 26 grains
of oxygen, nearly 19 and 7 as related above;
of which the 19 grains cannot be correctly
estimated by reason of the uncertainty as to
the real quantity of oxide formed during the
operation: there might be some left adhering
to the retort, or on the other hand there might
be more than the due share, derived from
former experiments.   Supposing then, that
26 grains of oygen were extracted from the
nitric acid, the remaining acid would require
the same to be added to re-form the nitric;
but by the experiments with oxymuriate of
lime it seemed to require 32 grains of oxygen.   This difference wants an explanation;
I believe the greater error must belong to
the 26 grains; perhaps the truth might be

R r

322 APPENDIX.

approximated best by supposing both to be
30 grains.

When the liquid decomposed nitre is treated
with any acid, a gas is instantly expelled
which produces red fumes in the air; it is
pure nitrous gas, which joining with the oxygen of the atmosphere, generates nitrous acid
vapour. At the same time, no doubt, the
sub-nitrous acid is disengaged from the potash, but that part of it which is real nitrous acid (1 atom azote to 2 of oxygen) is
retained by the water, whilst the other part,
(1 atom azote and 1 of oxygen) assumes the
gaseous form. In order to be satisfied respecting this point, I made several experiments with the liquid over mercury : taking
a given portion of the liquid, and sending it
to the top of a graduated tube filled with
mercury, I passed up as much muriatic acid
as was sufficient to engage the potash ; immediately there was a disengagement of
nitrous gas and carbonic acid gas, and afterwards a slow evolution of gas, evidently arising from the liquid in contact with the mercury. Wishing to ascertain the quantities, I
sent up 25 grain measures of liquid, and to
that nearly half its bulk of muriatic acid ; in
2 or 3 minutes there was,

1.1 cubic inch of gas.	H.	M.
1.4 .... in 0		45
1.5 1		5
1.7 2		45
1.75 7		45
1.78 9		45

The gas was washed in lime water, and lost .33 parts of an inch of carbonic acid ; the rest, 1.45 cubic inch, was nitrous gas. It is obvious that ½ of the nitrous gas, together with the carbonic acid, was liberated instantly ; the rest of the nitrous gas was due to the nitrous acid, slowly acting upon the mercury. At the end of the process, there was a little black oxide floating upon the mercury. Calculating from this, the whole quantity of nitrous gas would be 31 or 32 grains, whereas it ought to have been 48 grains to constitute 62 of sub-nitrous acid. It is probable that whilst a portion of the subnitrous acid is oxidizing the mercury, another portion may be forming nitric acid and dissolving the oxide.

From some trials, I have reason to think that even carbonic acid will expel nitrous gas from the liquid sub-nitrite of potash.

In the essay of Dr. Henry, already alluded to, published in the 4th Vol. of the Manchester Society's Memoirs, a new and interesting discovery is made ; namely, that a mixture

of nitrous and olefiant gases, though not explosive by an electric spark, may still be
exploded by the more powerful impetus of a
shock from a charged jar. Dr. Henry has
adduced the results obtained in this way, as
corroboratory of those which shew the constitution of nitrous gas to be 1 volume of azote
and 1 of oxygen united to form 2 volumes of
nitrous gas. (See page 507 of the Memoirs.)

Some time ago in repeating these experiments of Dr. Henry, I found some extraordinary circumstances attending them. After
determining that 1 volume of olefiant gas
may be fired with from 6 to 10 volumes of
nitrous, I found a shock from a jar sometimes
inadequate to fire the mixture, which, however, when repeated a second or third time,
succeeded. This is not a novelty; for, mixtures of olefiant gas as well as other gases and
vapours, with a minimum of oxygen, frequently require several sparks before the
explosion : but this case occurs at times with
nitrous and olefiant gas, when they are mixed
in the most favourable proportions for exploding. The most remarkable circumstance,
however, was, that when a phial was filled
with the mixture of the two gases in the proportion of 1 volume olefiant to 6 or 7 nitrous,
(exclusive of small portions of azote), the

decomposition of the nitrous gas and the combustion of the olefiant were scarcely ever perfect ; and what increased the perplexity more,
was, the results obtained from the same mixture scarcely ever agreed one with the other.
After about 30 experiments, I was inclined to
adopt the conclusion, that the uncertainty was
occasioned by the oblong form of the eudiometer.   The spark or shock, in my eudiometer, is imparted at one extremity of a column
of air, which is often 10 times as much in
length as in diameter: it mostly was found
that the larger the quantity of mixture exploded at once, the more imperfect and in-
complete was the combustion.   I imagine
the intensity of heat is not sufficient to carry
on the combustion through the length of the
column, owing, perhaps, to the cooling power
of the sides of the tube.   Hence it was, I apprehend, that in one or two instances, when
a small quantity of gas was used, I got
nearly complete results, as Dr. Henry reports
his ; but in the majority both gases were
found in the residue after the explosion.

In pursuing this enquiry into the decomposition of nitrous gas by combustible gases, I
found that it might be effected by any combustible gas or vapour: at least it succeeded
in all I tried.   The method I pursued, and

which was suggested by the known properties of phosphuretted hydrogen, is this: it
has been shewn (page 181) that a mixture of
phosphuretted hydrogen and nitrous gas exploded by an electric spark, the former gas
being completely burned in case the proportions are duly adjusted ; now, it occurred to
me, that as the above combustible gas is usually a mixture of pure phosphuretted hydrogen and of hydrogen, and that the latter of
these is also burned as well as the former, the
effect must be produced through the heat
occasioned by the combustion of the former.
Having some old phosphuretted hydrogen by
me, at the time, which on examination, I
found to be 91 per cent. combustible gas, and
9 azote ; and the 91 combined with 156 of
oxygen, consequently was 74 pure, and 17
hydrogen ; I tried this mixture with nitrous
gas, when it exploded by the spark, as usual ;
but on trying it with an excess or defect of
nitrous gas, the spark was inefficient, but the
shock instantly fired the mixture. As there
did not appear to be any of the pure hydrogen
left unburned in these experiments, I proceeded to mix the old phosphuretted hydrogen with hydrogen ; and then this new mixture with nitrous gas. The first experiment
was made with 4 parts of old phosphuretted

hydrogen + 16 hydrogen + 36 nitrous gas
= 56 total. On this mixture the spark, of
course, had no effect ; but it exploded the first
trial by the jar, and left 20 measures, of which
2 were found to be oxygen, and the rest azote.
This experiment succeeding so well, I next
tried mixtures of phosphuretted hydrogen,
with carbonic oxide, carburetted hydrogen,
and ether vapour successively, along with
nitrous gas ; and found that all these mixtures refused combustion by the spark, but
were instantly exploded by the shock, yielding carbonic acid and water, the same as if
the combustion had been effected by free oxygen. In some instances the combustion was
complete, leaving neither combustible gas
nor nitrous gas ; but generally there was a
residue of one or both of the gases.

From these experiments it may be concluded that the heat, produced by the combustion of phosphuretted hydrogen and nitrous
gas or oxygen gas, disposes other gases, accidentally in the mixture, to chemical changes.
In conformity with this view, I mixed phosphuretted hydrogen and oxygen, in the proportion of mutual saturation ; and taking a
small proportion of this mixture, and as much
ammoniacal gas as would saturate the phosphoric acid to be formed, I found that caus-

ing an explosion over mercury, the phosphoric acid combined with the ammonia, and nearly the whole gas disappeared. In this case, the heat was not sufficient to decompose the ammonia. But in another experiment, with a portion of the same explosive mixture and a less proportion of ammonia, after the firing a residue of azote and hydrogen was found, amounting nearly to the quantity due from the decomposition of the ammonia. Here the heat produced, had evidently decomposed the ammonia.

ON AMMONIA.

The constitution of ammonia still remains undecided. The latest experiments on this article are those of Dr. Henry, in his essay on the analysis of the compounds of nitrogen. (Memoirs of the Manchester Society, vol 4, 1824.) By electrifying ammoniacal gas over mercury, as carefully as could be devised, Dr. Henry found results as under:

1st experiment	44	measures became	88+
2d	157		320
3d	60		122
4th	120		240

The evolved gases carefully analysed by combustion with oxygen, were found to consist of

3 volumes of hydrogen and 1 of azote. The analysis of ammonia was also effected by exploding it with nitrous oxide, with the requisite precautions. The results confirmed the previous ones by electricity, both in regard to doubling the volume of ammonia, and establishing the ratio of 3 to 1 in the volume of hydrogen and azote.—These experiments are highly interesting as far as regards the question of ammonia, as they exhibit the latest investigations of one who has previously shewn uncommon skill and perseverance in this kind of analysis. (See Philos. Transact. 1809, &c.)

Dr. Henry's analysis of ammonia, in 1809, has been adverted to in our article on the subject, vol. 1, page 429. The results of that Essay are given in a tabular form; and the mean of six experiments was nearly as we have stated, namely, that ammonia consists of $27\frac{1}{4}$ measures of azote, and $72\frac{3}{4}$ hydrogen. To this it may be proper to add, that the two extremes were, 26.1 azote and 73.9 hydrogen, and 28.2 azote with 71.8 hydrogen; also that a small error has crept into the table, which being corrected, the average results are reduced to 27 and 73, very nearly. Subsequently, both Dr. Henry and Sir H. Davy concurred in assigning 26 and 74 for

S s

the most approximating numbers. (See
Nicholson's Journal, 25, page 153). The
true quantity of gases procured by the decomposition of ammoniacal gas by electricity,
was concluded by both these authorities, to be
180 for each 100 of ammonia, when the requisite precautions were taken, as we have
related in vol. 1.

From what is stated above, it is evident the
subject is one which requires extraordinary
skill and attention. This I can attest from
my own experience, which has been frequently renewed and varied; but the results
have not been sufficiently accordant to yield
me satisfaction.

About ten years ago, I made several
experiments on the decomposition of ammonia, which, though they are not convincing,
deserve, perhaps, to be recorded in their results.—Some more recent experiments are
incorporated with them.

Decomposition of ammonia by nitrous oxide.
—I made many experiments, by exploding
mixtures of nitrous oxide and ammoniacal
gases over mercury. The excess of gas was
mostly on the side of ammonia, but the proportions were varied in the different experiments, from 10 vol. nitrous oxide to 11 ammo-

nia or to 5, which are about the extremes capable of being fired by the electric spark.

When 10 parts nitrous oxide and 5 of ammonia are exploded over mercury, the residuary gas contains some free oxygen and some nitrous acid derived from the decomposition of the excess of nitrous oxide used ; with 6 parts of ammonia there is rarely any free oxygen. When 10 parts of nitrous oxide, and 7 of ammonia are fired, I never found any free oxygen or hydrogen; but when the ammonia is at or near 8 parts, I find from $\frac{1}{20}$ to $\frac{1}{10}$ of the hydrogen from the ammonia in the residuary gases. The two gases appear to be completely decomposed; the oxygen of the nitrous oxide, as far as it can, unites with the hydrogen of the ammonia, without forming any portion of nitrous acid or of free oxygen, and the residue contains the azote of both gases, and the unburnt hydrogen from the ammonia, as Dr. Henry first observed. This continues to be the case till the ammonia becomes 11 parts, when the hydrogen amounts to about $\frac{1}{3}$ of the whole quantity which the ammonia yields.

From the above it would seem that the proportions for mutual saturation must be 10 nitrous oxide with from 7 to 8 parts of ammonia. This agrees with the deduction in

Dr. Henry's first essay that 13 nitrous oxide require 10 of ammonia ; or that 10 require 7.7 : but according to the theory of volumes 10 would require $6\frac{2}{3}$; and Dr. Henry recommends in his late essay 10 nitrous oxide to 7.7 or $8\frac{1}{3}$ parts of ammonia, in order to secure a small excess of the last, and consequently some free hydrogen after the explosion. The former of these proportions would have nearly $\frac{1}{7}$ of the residue hydrogen, and the latter nearly $\frac{1}{5}$, supposing the gases pure originally. This gives more hydrogen than I have ever found ; but the azote in my experience nearly agrees with the doctrine of multiple volumes.

Decomposition of ammonia by nitrous gas. — About 30 experiments carefully made on mixtures of nitrous gas and ammoniacal gas gave very discordant results. At one time 10 parts nitrous gas with 14 ammonia gave $\frac{1}{3}$ of hydrogen in excess, and another time 10 nitrous with 12 ammonia gave excess of hydrogen $= \frac{9}{20}$; generally 10 parts with 6 or less gave oxygen, and 10 with 8 or more gave hydrogen in the residue.

Decomposition of ammonia by oxygen. — The limiting proportions of oxygen and ammonia which I have fired, are 10 oxygen to 4 ammonia for the minimum, and 10 oxygen to 22 ammonia for the maximum. When 10 oxygen were fired with 4 ammonia, there were $\frac{25}{37}$ of the oxygen left, and there was a deficiency of azote amounting to $\frac{1}{12}$ of what was expected from the ammonia, owing no doubt to nitrous acid generated by the explosion. When 10 oxygen to 1.8, or from that to 2.2 ammonia are used, there is a surplus of about $\frac{1}{4}$ or $\frac{1}{3}$ of the hydrogen contained in the ammonia, left in the residue of the gas. When the ammonia is between 13 and 14 there is usually a trace of oxygen or hydrogen as it approaches either of these limits. By the theory of volumes, 10 oxygen should saturate $13\frac{1}{3}$ of ammoniacal gas. I have not any instance of hydrogen being left when 14 ammonia were used, though there ought to be $\frac{1}{10}$ of the whole left; and much smaller quantities than that are appreciable by well known methods. The azote resulting from the decomposition of ammonia is usually very nearly $\frac{1}{2}$ the volume of the ammonia.

On the whole the results from firing ammonia and oxygen gas appear to me more satisfactory than those obtained from nitrous oxide and nitrous gas, as they are more simple and less perplexed with any theoretic views.

It may be proper to remind the reader that when we speak of 10 parts of one gas uniting

with 8, 10, or more, of another in the above
and other cases, it is to be understood of
gases *absolutely pure ;* not that we ever obtain
them in that state, but approximating as near
as we can to it, we mix given portions of
such gases as we can obtain, and then in our
calculations of results deduct for the impurities.

One source of uncertainty in these experiments on firing mixtures of ammonia, is that
the real quantity of ammoniacal gas operated
upon is not known. If a certain measure of
ammonia be transferred through mercury
ever so dry, some portion of it gets entangled
in the mercury, and 100 measures become
perhaps 95: now in the explosion it is a question whether any part of the 5 measures absorbed is decomposed. I have marked this
attentively, and am persuaded that generally
speaking, little if any of that portion is decomposed; but some trace of it appears mostly
afterwards in the residue as it is liberated from
the pressure of its own kind of gas, and hence
easily rises into the gaseous mixture. Notwithstanding, when the loss of gas by trans-
fer amounts to 10 or 20 per cent, I have reason to believe that some part of it suffers
combustion occasionally.

Volume of gases from the decomposition of ammonia.—It has been observed (vol. 1. Ammonia) that Sir H. Davy obtained 180 measures of gases, by means of electricity, from 100 of ammonia as the maximum when the operation was performed with great care, and Dr. Henry in like circumstances, produced 181, whilst I found 187 measures; since that, as has been related, Dr. Henry has found 200 measures. It is not easy to account for these differences; I am inclined to the opinion that the volume of gases is very nearly doubled, but probably rather less than more. I find the experiments on the rapid combustion of ammonia agree best with that opinion.

Decomposition of ammonia by a red heat.— A short time since I repeated the decomposition of ammonia by passing the gas through a red hot copper tube. The proportion of azote to hydrogen, due allowance being made for a minute portion of atmospheric air, was upon the average of a number of experiments, 26 of the former to 74 of the latter.

Decomposition of ammonia by oxymuriatic acid.—I have made several experiments on this mode of decomposition since the results published in vol. 1, page 435. It is well known that a solution of oxymuriate of lime decomposes ammoniacal salts; water and muriatic acid are produced, azote liberated, and the acid previously combined with the ammonia is evolved. But this is not all; an excessively pungent gas or perhaps vapour is produced, exciting sneezing, and inducing catarrh; the constitution of this vapour is not well understood; it is never formed, as far as I know, without the presence of both oxymuriatic acid and ammonia. The results of such mixtures are of course complicated and likely to be unsatisfactory; it may notwithstanding be useful to relate some of them.

When clear oxymuriate of lime solution, and a salt of ammonia are mixed together with a little excess of oxymuriate, the ammonia is mostly decomposed, the oxymuriate being converted into muriate of lime by the hydrogen of the ammonia, whilst the azote is evolved, and the acid previously combined with the ammonia is liberated; hence oxymuriatic acid gas is also liberated along with the azote; and it is required to be taken out before the azote can be estimated. This circumstance may be obviated by previously adding the requisite quantity of pure potash or soda, to engage the acid, or by leaving a little undissolved lime in the oxymuriatic solution. I could never obtain a volume of azote equal to half that of the am-

monia (supposed to be in a gaseous state)
though it is universally allowed not to be less
than that, if the whole of the azote be evolved;
on one occasion only I got so much as $\frac{14}{15}$ of
that quantity. The residue of liquid has the
extremely pungent smell; but the azotic gas
after passing through pure water has no smell.
When this experiment is made over mercury,
the oxymuriatic acid acts upon it, and hence
the excess of oxymuriate should be such as to
leave a portion of that undecomposed at the
conclusion.

When the object is to ascertain the hydrogen in ammonia, a portion of salt known to
contain a given weight of ammonia is to be
treated with oxymuriate of lime solution, the
strength of which is accurately determined by
means of green sulphate of iron, or otherwise.
The ammoniacal salt in solution is then to be
mixed with a moderate redundance of the
oxymuriate liquid, and with a few drops of
caustic potash, and the mixture must be repeatedly agitated for some time. At length
the liquid must be tested by the green sulphate of iron, and hence the quantity of aci
spent upon the ammonia will be determined.
I have mostly found the hydrogen this way
below the common estimate, allowing the
ammoniacal salts to be correctly determined.

T t

SULPHURET OF CARBON.

Since the article at page 462, vol. 1, was written, an excellent essay on the sulphuret of carbon has been published in the Philosophical Transactions, (1813) by Professor Berzelius and Dr. Marcet. After an extensive series of experiments, they infer the atom of the sulphuret to consist of 2 atoms sulphur and 1 of carbon. The investigation did not seem to warrant their including hydrogen in the atom. I have made several experiments on the combustion of the vapour of sulphuret of carbon in oxygen gas by electricity. My method generally was, to vapourize a given portion of atmospheric air over mercury, taking care that the vapour was below the maximum for the temperature; this is easily effected by putting the liquid into a phial of air, drop by drop, and inverting it over mercury till the liquid is evaporated. This vapourized air, I find may be transferred through mercury with very little loss, and even through water several times, without a total condensation of the vapour. The vapour of ether is much more condensible by water than that of sulphuret of carbon. A given portion of this vapourized air is to be mixed with oxygen gas, in Volta's eudiometer, and then exploded by the electric spark over mercury. One volume of vapour combines with nearly $3\frac{1}{2}$ of oxygen, and therefore requires 4 or 5 times its bulk of that gas before firing, in order that the combustion may be complete. The results of the combustion are carbonic acid and sulphurous acid; and I suspect a small portion of water; though Professor Berzelius and Dr. Marcet could not detect any.

By evaporating a given weight of the sulphuret of carbon, in a given volume of atmospheric air, at the temperature of 60°, I find the specific gravity of the vapour to be 2.75 nearly, air being 1. Now if we assume the atom of vapour to be nearly of the same volume as that of hydrogen, and to consist of 1 atom hydrogen, 2 sulphur, and 1 carbon, it will require 7 atoms of oxygen to form water, sulphurous acid, and carbonic acid, which will accord very well with my experience. When vapourized hydrogen gas is electrified for some time, there is no change of volume, though there is some appearance of decomposition. Probably the hydrogen of the sulphuret is liberated. It is difficult to conceive how so volatile a liquid as the one in question, could be constituted out of sulphur and carbon without the addition of hydrogen.

POTASSIUM, SODIUM, &c.

Two views of the nature of these bodies
have been given in vol. 1, (see pages 260, and
484, &c). In the former they are considered
as simple metals; in the latter, as compound
bodies resulting from the abstraction of oxygen from the hydrates of potash and soda;
or as being constituted of 1 atom of hydrogen
united to 1 atom of pure potash or soda respectively. Those who have had the most experience on these elements, Sir H. Davy,
and M. M. Gay Lussac and Thenard, seem
now to concur in the former view, and it has
been adopted by most chemists. Part of the
objections which we made to this view have
been obviated, it should seem, by establishing
the fact, that oxymuriatic gas and hydrogen
gas united, form muriatic acid gas. There
are still, however, difficulties to remove before
this view can be considered perfectly satisfactory; but they are not greater perhaps
than would attach to any other explanation
of the facts connected with the subject. Besides potassium and sodium, experience as
well as analogy would seem to render probable, if not to establish, the existence of barium, strontium, and calcium as metals, of
which barytes, strontites, and lime are the

protoxides, as potash and soda are of the other two metals ; (other oxides of potassium and sodium are stated, see page 55—57); barium has a deutoxide, and probably calcium likewise. The rest of the earths, as magnesia, alumine, silex, &c. are by analogy considered by most chemists as oxides of particular metals, but the proportions of their elements have not been determined.

ALUM.

At page 531, vol. 1, we have given the constitution of this important salt, as under : since that time Mr. R. Phillips has announced another view of it ; and Dr. Thomson has published one differing from both of these. They are as follow :

Dalton— 1 atom sulphate of potash.
4 atoms sulphate of alumine.
30 atoms water.

Phillips— 1 atom bi-sulphate of potash.
2 atoms sulphate of alumine.
22 atoms water.

Thomson— 1 atom sulphate of potash.
3 atoms sulphate of alumine.
25 atoms water.

Notwithstanding these differences, there is a near approximation in all three, in regard to the quantities of acid, alumine, potash, and water in the salt. This is accounted for partly in the different relative weights of the atoms, as estimated by the different analysts, but chiefly in that of alumine.

Some very curious results occurred to me about 10 years ago in analysing alum; they were new to me, but I have since found they had been previously discovered by Scheele. (See his essay on silex, clay, and alum, 1776.) As his observations are not to be found in any of our elementary books that I have seen, I shall give the particulars of my own experiments here.

I take 24 grains of alum and dissolve them in water; of these 8 grains may be allowed for sulphuric acid, $\frac{1}{5}$ of which $= 1.6$ grain $= 1.1$ grain of lime $= 880$ grains of lime-water, such as I commonly use. To the solution of alum I put 880 grains of lime-water; a slight precipitate appears which soon becomes redissolved almost completely. The liquid is then acid by the colour test.

To this liquid I put 880 more of lime-water, and agitate; a copious precipitate appears and continues; after subsidence the clear liquid is still acid by the colour test.

Another 880 grains are added, and the whole is then well agitated ; the agitation is repeated two or three times after the precipitate has partly subsided, so as to diffuse it equally again through the liquid; finally, the clear liquid is found to be neutral by the colour test, and to contain no alumine; for, lime-water produces no precipitate when poured into it.

Another 880 grains being added, and the whole stirred well, the clear liquid after the subsidence of the precipitate is still neutral by the colour test.

The fifth portion of 880 grains being then added, and the mixture well agitated, a considerable portion of the precipitate will evidently disappear, and the mixture become semitransparent ; after a time the clear supernatant liquid is found strongly alkaline ; a little of it touched with an acid becomes milky, and adding more acid clears it again. The liquid is now 1.0025 sp. gr., or a little heavier than lime-water.

The sixth portion of 880 grains being now added to the whole mixture, and agitated, the precipitate rather diminishes, and an increase of specific gravity takes place in the liquid ; it is now 1.003.

The seventh and last portion of 880 grains being added to the mixture, and agitation being continued for some time, a dense bulky precipitate is formed, which falls with great celerity, carrying with it the greatest part of the acid, the alumine and the lime, and leaving the liquid of the sp. gr. 1.0012. It is a subsulphate into which acid, potash, lime and alumine enter, as will be shewn.

These phenomena appear to me to be best explained by adopting a constitution of alum, such as to make it consist of 1 atom bisulphate of potash and 3 atoms of sulphate of alumine; after which the following explanation will apply.

The first portion of lime-water saturates the excess of acid.

The second portion throws down a correspondent portion of alumine. The clear liquid is acid, because it contains sulphate of alumine, which is essentially acid by the colour test, because alumine is not an alkaline element.

The third portion throws down another portion or atom of alumine; but by continued agitation the two atoms of alumine liberated, join the remaining atom of sulphate of alumine, and the whole compound falls down, being then the common subsulphate of alum.

Hence the liquid, containing nothing but sulphate of lime and sulphate of potash, is neutral by the test, and yields no alumine by the addition of lime-water.

The fourth portion of lime-water being put in and duly agitated, the atom of sulphuric acid is drawn from the subsulphate to join the lime, and then the floating subsulphate of alumine becomes pure alumine, and the clear liquor is still neutral.

The fifth portion of lime-water tries to decompose the sulphate of potash, but is unable of itself; however, the floating alumine assits it, and by double affinity the potash leaves the acid to join the alumine, and the lime takes the acid. Hence as $\frac{1}{3}$ of the alumine enters into solution with the potash, the precipitate is less copious, and the liquid is alkaline; a small portion of acid put into the clear liquid engages the potash, and liberates the alumine, but a larger portion redissolves the alumine also.

The sixth portion of lime-water seems to complete the effect which the fifth commences, and hence the density of the liquid increases, whilst the precipitate rather diminishes.

The seventh portion of lime, together with the sixth, after due agitation and some time, unite the lime with the alumine, one atom of

U u

each, and form a precipitate which would fall together, were no other compound present, as I found, and Scheele before me; but if sulphate of lime be present, each compound atom of lime and alumine, unites with one of sulphate of lime, and the whole descends together, forming a subsulphate resembling that of alum, only two atoms of lime are found as substitutes for two atoms of alumine. This subsalt is very little soluble in water.

According to this view, if 2 atoms of alum were decomposed, 4 atoms of subsulphate would be formed, each consisting of 1 acid, 2 lime, and 1 alumine; also 2 compound atoms of potash and alumine, and 6 atoms sulphate of lime. But in the final arrangement, it would seem, that 2 atoms of sulphate of lime are again decomposed, and sulphate of potash formed, the 2 atoms of lime combining with the 2 of alumine, and then two more atoms of subsulphate are formed, and the final arrangement is 6 atoms subsulphate precipitated, and 2 atoms sulphate of potash, and 2 sulphate of lime remain in solution.

The facts above stated appear to me to place the constitution of alum in a clearer point of view than any other I have seen. They make no difference in the weights of the several elements in 100 grains of the salt,

from what we have given in Vol. 1 ; only the
weight of the atom of alumine is here taken
to be 20 instead of 15, and we have 3 atoms
of it in 1 of alum, instead of 4, as in the
former account.

ON THE PRINCIPLES OF THE ATOMIC
SYSTEM OF CHEMISTRY.

It is generally allowed that the great objects of the atomic system are, 1st to determine the relative weights of the simple elements ; and 2d to determine the *number,* and
consequently the weight, of simple elements
that enter into combination to form compound
elements. The greatest *desideratum* at the
present time is the exact relative weight of
the element hydrogen. The small weight of
100 cubic inches of hydrogen gas, the important modifications of that weight by even
very minute quantities of common air and
aqueous vapour, and the difficulties in ascertaining the proportions of air and vapour in
regard to hydrogen, are circumstances sufficient to make one distrust results obtained by
the most expert and scientific operator. The
specific gravity of hydrogen gas was formerly
estimated at $\frac{1}{10}$ that of common air; it de-
scended to $\frac{1}{12.5}$, which is the ratio we adopted

in the Table at the end of Vol. 1. It is now commonly taken to be $\frac{1}{14.5}$, and whether it may not in the sequel be found to be $\frac{1}{16.5}$ is more than any one at present, I believe, has sufficient data to determine. The other factitious gases have mostly undergone some material alterations in their specific gravities in the last twenty years, several of which I have no doubt are improvements : but when we see these specific gravities extended to the 3rd, 4th, and 5th places of decimals, it appears to me to require a credit far greater than any one of us is entitled to. In the mean time, it may be thought a fortunate circumstance, that the weight of common air has undergone no change for the last thirty or forty years; 100 cubic inches being estimated to weigh 30.5 grains at the temperature of 60°, and pressure of 30 inches of mercury : (whether this is exclusive of the moisture I do not recollect.) It is also a fortunate circumstance, (provided it be correct) that this weight is nearly free from decimal figures. I may be allowed to add, that according to my experience, the weight of 100 cubic inches of air is more nearly 31 grains than 30.5. I apprehend these observations are sufficient to shew that something more remains to be done before we obtain a tolerably correct table of the specific gravities of gases; the importance of this object can not be too highly estimated.

The combinations of gases in equal volumes, and in multiple volumes, is naturally connected with this subject. The cases of this kind, or at least approximations to them, frequently occur; but no principle has yet been suggested to account for the phenomena; till that is done I think we ought to investigate the facts with great care, and not suffer ourselves to be led to adopt these analogies till some reason can be discovered for them.

The 2d object of the atomic theory, namely that of investigating the *number* of atoms in the respective compounds, appears to me to have been little understood, even by some who have undertaken to expound the principles of the theory.

When two bodies, A and B, combine in multiple proportions; for instance, 10 parts of A combine with 7 of B, to form one compound, and with 14 to form another, we are directed by some authors to take the smallest combining proportion of one body as representative of the elementary particle or atom of that body. Now it must be obvious to any one of common reflection, that such a rule will be more frequently wrong than right. For, by the above rule, we must consider the first of the combinations as containing 1 atom of B, and the second as containing 2 atoms of B, with 1 atom or more of A; whereas it is equally probable by the same rule, that the compounds may be 2 atoms of A to 1 of B, and 1 atom of A to 1 of B respectively; for, the proportions being 10 A to 7 B, (or, which is the same ratio, 20 A to 14 B,) and 10 A to 14 B; it is clear by the rule, that when the numbers are thus stated, we must consider the former combination as composed of 2 atoms of A, and the latter of 1 atom of A, united to 1 or more of B. Thus there would be an *equal* chance for right or wrong. But it is possible that 10 of A, and 7 of B, may correspond to 1 atom A, and 2 atoms B; and then 10 of A, and 14 of B, must represent 1 atom A, and 4 atoms B. Thus it appears the rule will be more frequently wrong than right.

It is necessary not only to consider the combinations of A with B, but also those of A with C, D, E, &c.; as well as those of B with C, D, &c., before we can have good reason to be satisfied with our determinations as to the *number* of atoms which enter into the various compounds. Elements formed of azote and oxygen appear to contain portions of oxygen, as the numbers 1, 2, 3, 4, 5, successively, so as to make it highly improbable that the combinations can be effected in any other than one of two ways. But in deciding which of those two we ought to adopt, we have to examine not only the compositions and decompositions of the several compounds, of these two elements, but also compounds which each of them forms with other bodies. I have spent much time and labour upon these compounds, and upon others of the primary elements carbone, hydrogen, oxygen, and azote, which appear to me to be of the greatest importance in the atomic system; but it will be seen that I am not satisfied on this head, either by my own labour or that of others, chiefly through the want of an accurate knowledge of combining proportions.

NEW TABLE
OF THE RELATIVE WEIGHTS OF ATOMS.

At the close of the last volume, the weights of several principal chemical elements or atoms were given; but as several additions and alterations have been educed from subsequent experience, it has been judged expedient to present a reformed table of weights.

SIMPLE ELEMENTS.

	Weights.		Weights.
Hydrogen	1	Strontium	39
Azote	5±, or 10?	Antimony	40
Carbone	5.4	Iridium	42
Oxygen	7	Palladium	50
Phosphorus	9	Uranium	50, or 100?
Sulphur	13, or 14	Tin	52
Calcium	17?	Copper	56, or 28?
Sodium	21	Rhodium	56
Arsenic	21	Titanium	59?
Molybdenum	21, or 42?	Gold	60±
Cerium	22?	Barium	61
Iron	25	Bismuth	62
Manganese	25	Platina	73
Nickel	26	Tungsten	84, or 42?
Zinc	29	Silver	90
Tellurium	29, or 58?	Lead	90
Chromium	32	Columbium	107? 121?
Potassium	35	Mercury	167, or 84?
Cobalt	37		

SIMPLE OR COMPOUND?

	Weights.		Weights.
Fluoric Acid	10? 15?	Muriatic Acid Gas	30, or 31
Magnesia	17	Zircone	45
Alumine	20	Silex	45?
Glucine	23? 34?	Yttria	53? 36? 18?
Lime	24		
Oxymuriatic Acid (chlorine)	29 or 30		

COMPOUND ELEMENTS.

	Weights.
Ammonia	6 ? 12 ? 13?
Olefant Gas	6.4 ? 12.8?
Carburetted Hydrogen or Pond Gas	7.4
Water	8
Phosphuretted Hydrogen	10
Nitrous Gas	12, or 24?
Carbonic Oxide	12.4
Sulphuretted Hydrogen	15
Deutoxide of Hydrogen	15
Nitrous Oxide	17
Nitrous Acid	19, or 38?
Carbonic Acid	19.4
Sulphurous Oxide	21
Phosphoric Acid	23
Sulphurous Acid	28
Protoxide of Arsenic	28
Soda	28
Hydrate of Lime	32
Protoxide of Iron	32
Protoxide of Manganese	32
Protoxide of Nickel	33
Sulphuric Acid	35
Sulphuret of Arsenic (native)	35
Hydrate of Soda	36
Oxide of Zinc	36
Carbonate of Magnesia	36.4
Protosulphuret of Iron	39
Deutoxide of Manganese	39
Oxide of Chromium	39
Muriate of Magnesia	39
Protosulphuret of Nickel	40
Protosulphuret of Lime	41
Carbonate of Lime	43.4
Protoxide of Cobalt	44
Strontites	46
Muriate of Lime	46
Chromic Acid	46
Protoxide of Antimony	47

	Weights
Carbonate of Soda	47.4
Hydrate of Potash	50
Muriate of Soda	50
Sulphate of Magnesia	52
Sulphuret of Antimony	54
Sulphate of Alumine (simple)	55
Oxide of Palladium	57
Sulphate of Lime	59
Protoxide of Tin	59
Carbonate of Potash	61.4
Hydrosulphuret of Antimony	62
Nitrate of Magnesia	62
Sulphate of Soda	63
Protoxide of Copper	63
Muriate of Potash	64
Deutoxide of Tin	66
Protosulphuret of Tin	66
Oxide of Gold	67
Barytes	68
Muriate of Lime	69
Oxide of Bismuth	69
Deutoxide of Copper	70
Nitrate of Soda	73
Sulphuret of Gold	74
Protosulphuret of Bismuth	76
Sulphate of Potash	77
Oxide of Platina	80 ?
Nitrate of Potash	87
Carbonate of Barytes	87
Muriate of Barytes	90
Oxide of Silver	97
Protoxide of Lead	97
Minium	98
Sulphate of Barytes	103
Deutoxide of Lead	104
Protosulphurets of Lead and Silver	104
Nitrate of Barytes	113
Protoxide of Mercury	174?
Deutoxide of Mercury	181 ?
Protosulphuret of Mercury	181 ?
Alum	277

X v

ADDENDA.

——

STEEL.— Since writing the article at page 214, I have had an opportunity of analysing the crystalline steel, formed by Mr. Macintosh's process of cementation by means of coal gas. I dissolved 21 grains of this steel in sulphuric acid, with only a very slight excess of acid. The whole was dissolved except about $\frac{1}{10}$ of a grain of silvery-like particles. The gas obtained amounted to 29.6 cubic inches. It yielded no trace of carbonic acid. When fired with oxygen it yielded 3 per cent. upon the volume of hydrogen of carbonic acid; and this arose, as I ascertained, from the hydrogen containing 3 per cent of carburetted hydrogen gas: it contained no carbonic oxide. Supposing the carbone to have been combined with the iron, it would amount only to about $\frac{4}{5}$ of a grain, to 100 grains of iron. Whether such a quantity can be deemed an essential or an accidental ingredient of steel, may be a subject of consideration.

By a mistake of the Printer, the following paragraphs were omitted after page 308.

EXAMPLE.

According to the following values of the different specific gravities, (of the accuracy of some of which there may be doubts) and referring to my essay on oil gas (Manchester Memoirs, Vol. 4, new series, page 79,) we may take the oil gas, which, when the incombustible portion was abstracted would be nearly .812 sp. gravity, and

100 pure gas give 152 carb. acid and take 248 oxygen;

Here $w = 100$, $a = 152$, $g = 248$, $S = 1.458$, $f = .555$, $c = .972$ $s = .0694$ and $C = .812$. The value of u reduces to the following form;

$$u = \frac{4\ 7916\ a - 1\frac{2}{3}\ g + 1.875\ w - 6\ C\ w}{.625} = 24.5$$

hydrogen per cent. of pure combustible gas.

Hence we have 75.5 volumes left for the 3 other ingredients $= w$ of the formula; and abstracting 12 $+$ from the oxygen on account of the hydrogen, $g = 236 -$, and $a = 152$ as above.

$$\text{Whence } x = \text{Superolefiant} = 38\tfrac{1}{4}$$
$$y = \text{Carb. hydrg.} = 30.2$$
$$\text{and } z = \text{Carb. oxide} = \underline{7+}$$
$$75.5$$

These results differ considerably from those deduced in the above essay; probably in part from errors in the above estimates of the specific gravities of one or more of the gases.

EXPANSION OF LIQUIDS BY HEAT.

I am not aware of any particular labour that has recently been given to the enquiry how far pure liquids accord with each other in the law which I announced as derived from the experiments on water and mercury, and corroborated by those upon several other liquids. See Vol. 1, Table of temperature, page 14; also page 36, and following.

Perhaps all liquids should be considered as *pure* that are subject to uniform congelation at certain temperatures on the one hand, and on the other are capable of being distilled by heat without any alteration in their constitution. Water and mercury will rank in the first place; alcohol of .82 specific gravity and ether of .72; concentrated sulphuric

acid ; nitric acid of 1.42 specific gravity :
naphtha and oil of turpentine, &c. will probably be thought to claim the next place.
It is desirable that the temperatures at which
these liquids congeal should be ascertained ;
also whether any decomposition is effected
by the operation. If these expand proportionally to a scale of square numbers for
certain given equal or unequal intervals of
temperature, it may point out something relative to the collocation of the ultimate particles in liquids. The apparent coincidence
of this rate of expansion in liquids, with the
geometrical progressive force of steams or
vapours creates an additional interest. It
may be that most or all of these supposed
relations are accidental, and only approx-
imative like that of the rate of expansion of
air and mercury, between the temperatures
of $-40°$ and $212°$; but I cannot think this
probable. Even should they be only approximations, they are of sufficient utility to be
kept in view.

FINIS.

Printed by the Executors of S. Russell

BOOKS, ESSAYS, &c.,

PUBLISHED BY THE SAME AUTHOR.

Meteorological Observations and Essays.
4*s.* 8vo. 1793.

Elements of English Grammar: or, a new
System of Grammatical Instruction, for the use
of Schools and Academies. 2*s.* 6*d.* 12mo. 1801.

☞ *A few Copies of these Works may still be had of the London Booksellers.*

A New System of Chemical Philosophy
Part I. of Vol. I. 7*s.* 8vo. 1808.

Part II. ————10*s.* 6*d.* 8vo.—1810.

Published by G. Wilson, Bookseller, Essex-street. Strand, London.

*ESSAYS, by the same, in the Memoirs of the Literary
and Philosophical Society, Manchester.*

Vol. 5. Part 1 —Extraordinary facts relating to the vision
of colours.

Part 2.—Experiments and observations to determine
whether the quantity of rain and dew is equal to the
quantity of water carried off by the rivers, and raised by
evaporation ; with an enquiry into the origin of springs.

Experiments and observations on the power of fluids,
to conduct heat, with reference to Count Rumford's
seventh essay on the same subject.

Experiments and observations on the heat and cold
produced by the mechanical condensation and rarefaction
of air.

Experimental essays on the constitution of mixed
gases ; on the force of steam or vapour from water and
other liquids, in different temperatures, both in a Torri-
cellian vacuum, and in air ; on evaporation ; and on the
expansion of gases by heat.

ESSAYS, &c.

Meteorological observations made at Manchester, from 1793 to 1801.

Vol. 1. *Second series.*—Experimental enquiry into the proportions of the several gases or elastic fluids constituting the atmosphere.

On the tendency of elastic fluids to diffusion through each other.

On the absorption of gases by water and other liquids.

Remarks on Mr. Gough's two essays on the doctrine of mixed gases; and on Professor Schmidt's experiments on the expansion of dry and moist air by heat.

Vol. 2. On respiration and animal heat.

Vol. 3. Experiments and observations on phosphoric acid and on the salts denominated phosphates.

Experiments and observations on the combinations of carbonic acid and ammonia.

Remarks tending to facilitate the analysis of spring and mineral waters.

Memoir on sulphuric ether.

Observations on the barometer, thermometer, and rain, at Manchester, from 1794 to 1818 inclusive.

Vol. 4. On oil, and the gases obtained from it by heat.

Observations in Meteorology, particularly with regard to the dew-point, or quantity of vapour in the atmosphere; made on the mountains in the North of England.

On the saline impregnation of the rain which fell during the late storm, December 5th, 1822—with an appendix to the same.

On the nature and properties of indigo, with directions for the valuation of different samples.

In the Philosophical Transactions of the Royal Society.

On the Constitution of the Atmosphere.—1826.

In Mr. Nicholson's Philosophical Journal.

Vol. 5. (Quarto) On the constitution of mixed elastic fluids, and the atmosphere.—1801.

Vol. 3. (Octavo) On the theory of mixed gases.

5. On the zero of temperature.

6. Correction of a mistake in Dr. Kirwan's essay on the state of vapour in the atmosphere.

ESSAYS, &c.

8. On chemical affinity as applied to atmospheric air.

9. Observations on Mr. Gough's strictures on the theory of mixed gases.

10. Facts tending to decide at what point of temperature water possesses the greatest density.

12. Remarks on Count Rumford's experiments on the max. density of water.

13 & 14. On the max. density of water in reference to Dr. Hope's experiments.

28. On the signification of the word *particle* as used by chemists.

29 Observations on Dr. Bostock's review of the atomic principles of chemistry.

In Dr. Thomson's Annals of Philosophy.

VOL 1 & 2. On oxymuriate of lime.—1813.

3. Remarks on the essay of Dr. Berzelius, on the cause of chemical proportions.

7. Vindication of the theory of the absorption of gases by water, against the conclusions of M. De Saussure.

9 & 10. On the chemical compounds of azote and oxygen, and on ammonia.

11. On phosphuretted hydrogen.

12. On the combustion of alcohol, by the lamp without flame.

On the *vis viva*.

In Phillips's Annals of Philosophy.

VOL. 10. (new series). On the analysis of atmospheric air by hydrogen.

Printed in the United States
By Bookmasters